07 OCT 2003

KU-354-753

002

07 OCT 2003

Package & Label DESIGN

BEFORE OPENING THE PACKAGE & LABEL CD-ROM:

Be sure to turn off all virtual memory, and set your monitor's color depth to 256 color.
The CD-ROM will not run properly otherwise.

The CD-ROM runs exclusively from the CD and does not install to your hard drive.

The CD-ROM features selected images from Rockport Publishers' package and label archives.
To access these images click on the specific links:
Bags & Wraps, Beverages, Boxes, Consumer Products, Food, Hang Tags, Health, Beauty & Fashion,
Home & Office, Promotional, Toys, Games & Hobbies

Each area can be accessed from the main contents screen by clicking on the appropriate button.
Once in the project area, you have access to all of the
images by clicking on the numbers in the scrollable box at left.

To search for a specific design firm or client click the search button at the bottom of the project screen.

TO PLAY THE PACKAGE & LABEL CD-ROM:

Macintosh
1. Place the CD-ROM in your CD-ROM drive.
2. Open the CD-ROM by double-clicking the Package & Label icon.
3. Choose the appropriate pandlplayme file for your Macintosh: pandlplaymeMac or pandlplaymePPC.
4. Double-click the package and label icon.
5. Choose one of the four round buttons from the contents screen to access the ten main areas.
6. The help button on each screen will explain each screen's operation.

Windows/PC
1. Place the CD-ROM in your CD-ROM drive.
2. Choose pandl.exe and open this file.
3. Choose one of the four round buttons from the contents screen to access the ten main areas.
4. The help button on each screen will explain each screen's operation.

By opening the CD-ROM package, you accept the following: in no event will Rockport or its licensees be liable to you for any consequential, incidental, or special damages, or for any claim by any third party.

Copyright (c) 1997 by Rockport Publishers, Inc.

All rights reserved. No part of this book may be reproduced in any form without written permission of the copyright owners. All images in this book have been reproduced with the knowledge and prior consent of the artists concerned and no responsibility is accepted by producer, publisher, or printer for any infringement of copyright or otherwise, arising from the contents of this publication. Every effort has been made to ensure that credits accurately comply with information supplied.

First published in the United States of America by:
Rockport Publishers, Inc.
33 Commercial Street
Gloucester, Massachusetts 01930-5089
Telephone: (508) 282-9590
Fax: (508) 283-2742

Distributed to the book trade and art trade in the United States by:
North Light, an imprint of
F & W Publications
1507 Dana Avenue
Cincinnati, Ohio 45207
Telephone: (800) 289-0963

Other Distribution by:
Rockport Publishers, Inc.
Gloucester, Massachusetts 01930-5089

ISBN 1-56496-354-3

10 9 8 7 6 5 4 3 2

Designer: Monty Lewis
Front Cover Images: (Clockwise from top left) p.129, p.122, p.141, p.18, p.84
Back Cover Images: (Clockwise from top left) p.145, p.28, p.12, p.16

Printed in China

Package & Label
DESIGN

Rockport Publishers, Gloucester, Massachusetts

Package
& Label
DESIGN

introduction 6

CONTENTS

8   consumer products

health & fashion   44

66   food

beverages   102

138   promotion

index and directory   159

Package & Label DESIGN

INTRODUCTION

THE SUPERMARKET

I'm one of those freaks who hangs out at the Super Shop & Shop at 2:00 AM. This is generally a good time for me to leisurely feed any packaging fetishes without any annoying busy shoppers (who rarely share the same passions).

A huge fan of package design, I'd drive all night in search of a new soft drink with a bottle shaped like a lava lamp or a cookie with a cool logo. And I admit it: I'm guilty of buying stuff I don't need or couldn't possibly use just because the packaging is cool.

You can find me ready to drop hard, cold cash on a whim for anything uniquely creative—maybe it's Rainforest Crunch or a cereal or powdered drink that brings out the kid in me and has me cart-racing down the aisles, or maybe a simple combination of colors that intrigues me, like Harden & Huyse Chocolates.

This is the amazing part. Aside from a vague overture made in a magazine ad or commercial on MTV, no one has to push these items on me. There's no smiling salesperson—at this ungodly hour I'm more or less the only person in the store! So it comes down to me standing in the aisle with a carton, a label, a logo—the design. At its best, the design becomes more than the outer wrapper for the product, it becomes inseparable from it. For me, good design versus inseparable design is the difference between buying something once and being willing to go to jail for it.

But back to the Super Stop & Shop, where I assume I'll find most designers in the laundry detergent aisle when it comes time to design packaging for laundry detergent. This is a smart move; you certainly have to know what's out there. But what I also recommend, strangely enough, is that you check out the design solutions in produce. Yup, produce. Whether your assignment is a label for a coffee bag or a box for dog biscuits, don't pass up produce.

HERE'S WHY.

Like all good package and label design, a great deal of the packaging in produce is effective and highly functional. I had a professor who said that no one could package a banana but the banana. Let's face it, as good as we all think we might be, no one would take on the banana. Its natural packaging protects the product, keeps it fresh, and is easy to open. It's even color-coded for the various predilection of tastes: green, yellow, and brown. If a good looking bunch is eight-up and you only need six, no problem. Hot dog buns should be so flexible.

And consider the sheer diversity in the produce department. There's probably over six dozen fruits and vegetables, and no one product is cashing in on another's design. (We won't mention Oranges vs. Tangerines, 1966).

Have you ever noticed those kooky product items like the kiwi, starfruit, yucca, prickly pear, and kumquat? (Who even buys this stuff?) Still, every so often one of them hits big time. The kiwi is a case in point. Ten years ago you'd have to fly around the world for one. Now they sell them at the Piggly Wiggly and put them in your cereal at Denny's and Bickford's. These fruits and veggies will have you consider even the corniest (no pun intended) label or package design solution. It just might be what it takes to get folks to pick the fruits of your labor.

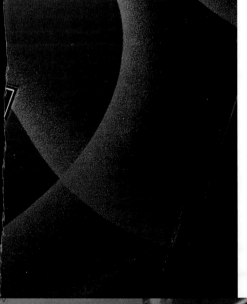

Now because I have an interest in tea, folks often point out that the tea section has recently become one of the most colorful. This is true. The packages are pretty day-glow these days, but the basic configuration, dimensions, and stock of the boxes are all nearly identical. It's getting monotonous. Even the typical Thursday, 6:30PM shopper would appreciate some novelty, if for nothing else, to differentiate the products as they sit on the shelves, which is paramount to good packaging design.

The point is that good packaging designs, nature-made or man-made, share pretty much the same principles. So absorb all you can in your late night supermarket run and in every aisle. Just be careful not to trip over the corrugated boxes and pallets of produce as the graveyard shift restocks the shelves. After all, that's packaging, too.

Jeff Piazza, Steep

Consumer Products

DESIGN FIRM Kristen Baumgartner Design
ART DIRECTOR/DESIGNER Kristen Baumgartner
ILLUSTRATOR Adam Cohen
CLIENT Imaginarium
PRODUCT Time Machines project in Science in a Shoebox series

The designer was asked to collage different clocks and illustrate some of the time projects that are part of the kit. The abstract background of colorful gears, oddly proportioned clocks, and time pieces from different eras are all layered together with night and daytime elements. Hardware used for this project was Power Computing's Power Tower 180. The illustration was created using Adobe Photoshop, and the cover was created using Adobe Illustrator.

DESIGN FIRM Mars Advertising
ART DIRECTOR Susan Sanderson
DESIGNER Michelle Uredevoogd
CLIENT Environmental Quality Co.
PRODUCT Software packaging
TECHNIQUE Screen, offset

The purpose of the software package is to assist the client in the tracking and government certification of hazardous and non-hazardous waste. It was important that the packaging visually communicate the company's continued position as an environmentally sensitive corporation. Production was done on Macintosh platform hardware using QuarkXPress and Adobe Illustrator and Photoshop.

DESIGN FIRM Mars Advertising
ART DIRECTOR Susan Sanderson
DESIGNER Michelle Uredevoogd
CLIENT Environmental Quality Co.
PRODUCT Software packaging
TECHNIQUE Offset

The purpose of the software package is to assist the client in the tracking and government certification of hazardous and non-hazardous waste. It was important that the packaging visually communicate the company's continued position as an environmentally sensitive corporation. Production was done on Macintosh platform hardware using QuarkXPress and Adobe Illustrator and Photoshop.

DESIGN FIRM Watts Graphic Design
ART DIRECTOR/DESIGNER Helen and Peter Watts
CLIENT Chris and Debra
PRODUCT Wedding invitation
TECHNIQUE Offset

Chris and Debra actually speak on this CD to invite their guests. It was unique and successful.

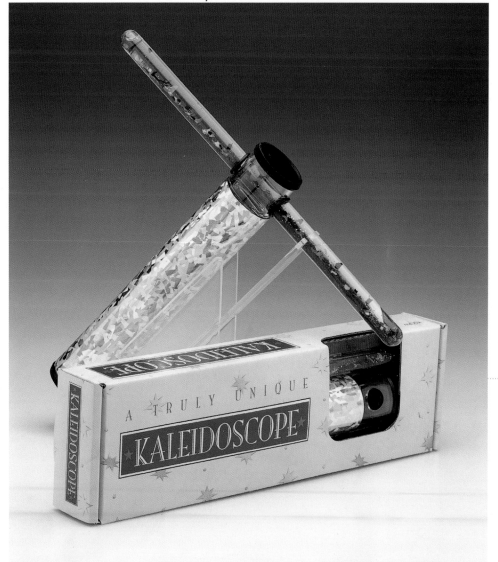

DESIGN FIRM Design Communications
ART DIRECTOR Curt Hamilton
DESIGNER Natalie Bianchi
CLIENT Quartet Manufacturing
PRODUCT Biggie Cakes
TECHNIQUE Offset

This client needed a new packaging design to appeal to a more sophisticated buyer of children's paints. Adobe Illustrator and QuarkXPress were used.

DESIGN FIRM Wood & Wood
ALL DESIGN Preston Wood
CLIENT Lyon Design
PRODUCT Kaleidoscope
TECHNIQUE Offset

For this kaleidoscope, a star fabric was scanned and colored in Adobe Photoshop. The file was then manipulated in QuarkXPress.

consumer products | 13

DESIGN FIRM Wood & Wood
ALL DESIGN Preston Wood
CLIENT Lyon Design
PRODUCT Paperweights
TECHNIQUE Offset

The Rude Cube and Crude Cube are paperweights that yell when you pick them up. They were designed with holographic coverings to attract attention. The type was done in Adobe Illustrator and QuarkXPress.

DESIGN FIRM Wood & Wood
ALL DESIGN Preston Wood
CLIENT Lyon Design
PRODUCT Executive toy
TECHNIQUE Offset

The photos were hand-tinted black and white. The type was done in Adobe Illustrator and QuarkXPress. These packages were made to look amusing and to stand out in the novelty category.

DESIGN FIRM Marsh, Inc.
ART DIRECTOR Greg Conyers
DESIGNER John Mikula, Greg Conyers
CLIENT Infotel
PRODUCT Computer hardware boxes
TECHNIQUE Offset

The firm's objective was to apply a new line identity to various computer hardware boxes and sleeves. QuarkXPress and Adobe Illustrator were used.

DESIGN FIRM
Marsh, Inc.
ART DIRECTOR
Greg Conyers, Ken Neiheisel
DESIGNER
Eileen Pieczonka, Greg Conyers
ILLUSTRATOR
Eileen Pieczonka
CLIENT
Infotel/Midwest Micro
PRODUCT
Computer hardware boxes
TECHNIQUE
Offset

The objective here was to apply a new line identity to various computer hardware boxes and sleeves. QuarkXPress and Adobe Illustrator were used.

consumer products | 15

DESIGN FIRM Smart Design Inc.
ART DIRECTOR Tucker Viemeister
DESIGNER Debbie Hahn, Stephanie Kim,
    and Nick Graham of Joe Boxer
CLIENT Timex/Joe Boxer
PRODUCT Watches
TECHNIQUE Offset

These round, cardboard, souvenir boxes with the smiley face logo serve as packaging as well as in-store displays. The copy inside of the box reminds the buyer that "this is not underwear." Humorous instructions are tucked inside along with tissue paper (printed with happy adjectives) to keep the watches from banging around. The whole package is made of recyclable paper.

DESIGN FIRM CommuniQué Marketing
ART DIRECTOR Gia Owens
DESIGNER Gia Owens
ILLUSTRATOR Andrew Portwood
CLIENT Carpenter Co.
PRODUCT Packaging for Dreamhugger series
TECHNIQUE Offset

This product was directed toward a younger market. For that reason, the designer chose energetic photographs to depict active lifestyles. The packaging also includes an enlarged photograph of the product to show what the foam actually looks like.

16 | Package & Label Design

DESIGN FIRM One Design
ALL DESIGN James Schenck
CLIENT Voice Navigator
PRODUCT Voice Navigator software

This spec job was done in order to create a more basic down-to-earth look, incorporating the microphone and wings. Unfortunately, this idea was never produced.

DESIGN FIRM Elizabeth Resnick Design Studio
ART DIRECTOR Robert Potts
DESIGNER/ILLUSTRATOR Elizabeth Resnick
CLIENT Chiron Diagnostics (formerly CIBA Corning)
PRODUCT Chemistry controls
TECHNIQUE Offset

Color-coded boxes and screw top vial caps minimize component mix-ups, and clear, flexible dispensing tips can be inserted into screw caps for ease of use. The designer used a full-bleed gray grid background with color-coded edges and white reversed-out and black graphics on a 24 pt. recyclable SBS box with auto-bottom. Soy inks, water-based adhesives, and UV-cured clear aqueous coatings were used.

consumer products | 17

DESIGN FIRM Pentagram Design Inc.
ART DIRECTOR Paula Scher
DESIGNER Lisa Mazur
ILLUSTRATOR Paula Scher, Lisa Mazur
CLIENT G.H. Bass & Co./Spirit of Maine
PRODUCT Specialty gift items
TECHNIQUE Offset

Spirit of Maine food, gardening, and gift products of New England are sold exclusively in G.H. Bass & Co. stores. Black-and-white maritime photographs accentuate the sub-brand's Yankee personality and origins.

DESIGN FIRM Clark Design
ART DIRECTOR Annemarie Clark
DESIGNER Thurlow Washam
PHOTOGRAPHER Geoffrey Nelson
CLIENT Oracle
PRODUCT Video
TECHNIQUE Offset

This video promotes the Oracle Channel, a division of Oracle Education. The packaging needed to use the same design that the firm had created for a brochure announcing the product.

18 | Package & Label Design

DESIGN FIRM Clark Design
ART DIRECTOR Annemarie Clark
DESIGNER Thurlow Washam
CLIENT Oracle
PRODUCT Software
TECHNIQUE Offset

This packaging design is used for an entire library of products that include manuals, video and software packaging, and an information binder with charts. Various job roles were coded on each piece so the users could quickly see which materials referred to them. The design was created in QuarkXPress and printed with three colors.

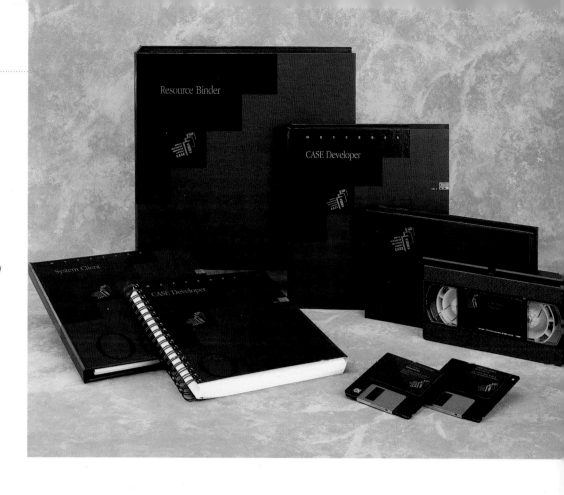

DESIGN FIRM Clark Design
ART DIRECTOR Annemarie Clark
DESIGNER/ILLUSTRATOR Thurlow Washam
CLIENT Oracle
PRODUCT Software
TECHNIQUE Offset

Oracle Education Division's computer-based software packaging uses the image of a column to signify the strength and foundation of education. A different part of the column was used for each piece. It was created in QuarkXPress, Adobe Illustrator, and 3-D software.

consumer products | 19

DESIGN FIRM Louise Fili Ltd.
ART DIRECTOR/DESIGNER Louise Fili
CLIENT Chronicle Books
PRODUCT Greeting cards
TECHNIQUE Offset

This series of cards was produced by Chronicle Books by using the designer's collection of French advertising fans from the twenties and thirties.

DESIGN FIRM Louise Fili Ltd.
ART DIRECTOR/DESIGNER Louise Fili
CLIENT El Paso Chile Co.
PRODUCT Floribunda
TECHNIQUE Offset

The challenge here was to find a way to package a kit of bulbs, soil, pot, and saucer. Louise Fili Ltd. came up with the name Floribunda and decided to make a tie-on tag that could include the copy.

20 | Package & Label Design

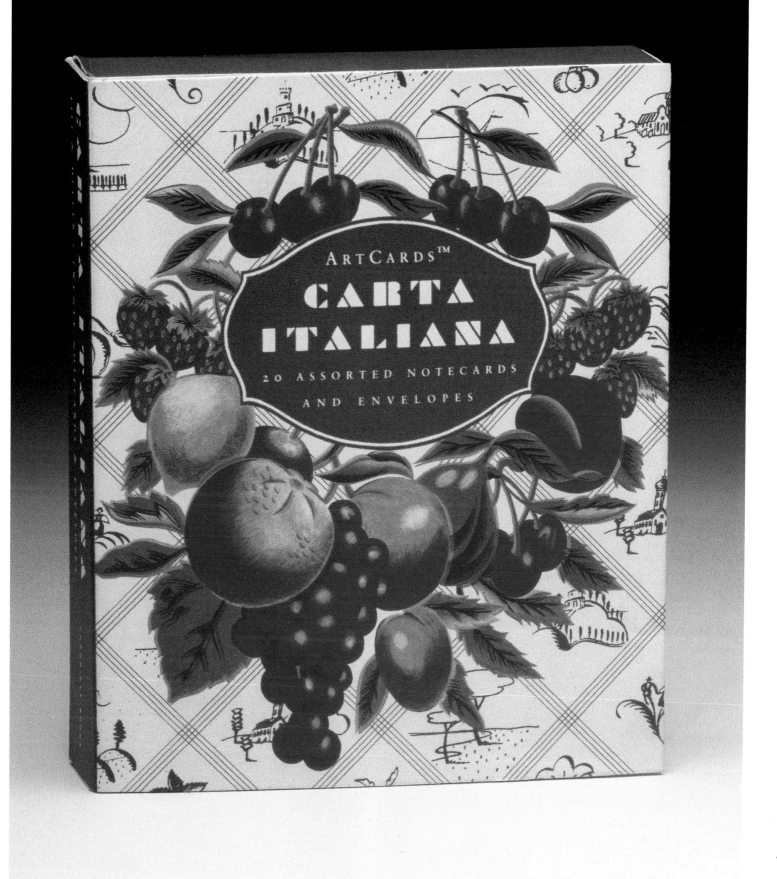

DESIGN FIRM Louise Fili Ltd.
ART DIRECTOR/DESIGNER Louise Fili
ILLUSTRATOR Melanie Parks
CLIENT Chronicle Books
PRODUCT Greeting cards
TECHNIQUE Offset

This box of greeting cards is based on the designer's collection of Italian orange wrappers. The illustration was done in gouache, and all type was done on the computer.

consumer products | 21

DESIGN FIRM CMA
ART DIRECTOR Bob Milz
DESIGNER Leon Alvarado, Trish Hill
ILLUSTRATOR Jeff Sanson
CLIENT BRIK Toy Co.
PRODUCT Rods&Pods
TECHNIQUE Airbrush illustration, offset

Rods&Pods is a new product engineered to fit with and connect to all the other leading building block systems. The cylindrical cardboard tube packages show a construction built with the contents of that size package, and additional illustrations around the tube emphasize the compatibility of this toy.

DESIGN FIRM Rickabaugh Graphics
ALL DESIGN Eric Rickabaugh
CLIENT Nationwide Insurance Company
PRODUCT Insurance software
TECHNIQUE Offset

Life Manager software is used by Nationwide Insurance agents to create insurance projections for clients. Because it helps agents "juggle" client facts and figures, Rickabaugh Graphics created the juggler image and applied it to labels, manuals, packaging, and announcements.

DESIGN FIRM Robilant & Associati
ART DIRECTOR Maurizio di Robilant
CLIENT Japan Tobacco Ltd.
PRODUCT Frontier cigarettes

To redefine the Frontier brand, both the logo and the packaging system had to be redesigned. After relaunch, the new image proved to be so successful that the brand increased its market share and added new items (100s and Menthol) to the Frontier line.

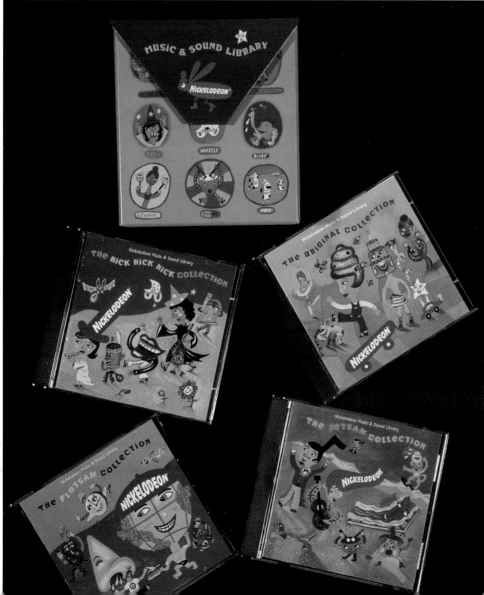

DESIGN FIRM Acme Creative Group
ART DIRECTOR Laurie Hinzman
DESIGNER Masako Moribayashi
ILLUSTRATOR Sara Schwartz
CLIENT Nickelodeon
PRODUCT Nickelodeon Music and Sound Library
TECHNIQUE Offset lithography

A library of music created for Nickelodeon is packaged in a box set holding four CDs. The artwork was inspired by the music and SFX on the CDs was inspired by Nickelodeon's personality.

consumer products | 23

DESIGN FIRM Stoltze Design
ART DIRECTOR Clifford Stoltze
DESIGNER Clifford Stoltze, Peter Farrell, Heather Kramer
ILLUSTRATOR 28 various artists
CLIENT Castle von Buhler Records
PRODUCT Anon art and music CD compilation
TECHNIQUE Offset

This package was designed to be a self-contained portfolio for two audio CDs and thirty-one postcards. It is an economical and visually unique alternative to a jewel case with a printed insert.

DESIGN FIRM Acme Creative Group
ART DIRECTOR Kenna Kay
DESIGNER/ILLUSTRATOR Melinda Beck
CLIENT Nick at Nite
PRODUCT Nick at Nite screen saver
TECHNIQUE Offset lithography

Nick at Nite created a screen saver as a direct mail piece for media buyers. The graphics tied the piece into a national ad trade campaign. This was Acme Creative Group's first multimedia piece, and it was very well received.

DESIGN FIRM Parham Santana Inc.
ART DIRECTOR Maruchi Santana
DESIGNER Millie Hsi
CLIENT Viva International Group
PRODUCT Sunglasses case
TECHNIQUE Four-color process over tin

A set of four collectible tins serve as a point-of-purchase display and also help to distinguish the client in a mature and crowded eyewear market.

DESIGN FIRM
Parham Santana Inc.
ART DIRECTOR
Maruchi Santana
DESIGNER/ILLUSTRATOR
Lori Reinig
CLIENT
Beacon Looms
PRODUCT
Tab top curtains
TECHNIQUE
Offset

"Elements" was a name and packaging concept conceived by Parham Santana Inc. for a graphic-driven curtain program. This packaging stands out in a marketplace where product photography is the norm.

consumer products | 25

DESIGN FIRM Carmichael Lynch
ALL DESIGN Pete Winecke
CLIENT West Publishing
PRODUCT Computer software
TECHNIQUE Silk screen

Creator, a software program for West Law Publishing, was a promotional piece sent to first-year law students to interest them in the West Law programs. The design had to appeal to a young consumer.

DESIGN FIRM Pedersen Gesk
ART DIRECTOR Rony Zibara
DESIGNER Rony Zibara, Kris Morgan, Mark Orton
CLIENT Anchor Hocking Plastics
PRODUCT Kitchen Essentials

The design strategy here was to introduce a revolutionary line of plastic containers aimed at "food preparation" oriented households—homes where people spend a fair amount of time in the kitchen. The design highlights the different storage solutions, and the color scheme is French Country.

DESIGN FIRM Pedersen Gesk
ART DIRECTOR Mitch Lindgren
DESIGNER Kris Morgan
CLIENT Target
PRODUCT Pet Essentials food and toys

This was the new private label design for Target Pet Essentials. The design appeals to the consumer's love for pets. The toys take the concept of the food packaging and add playfulness to the pets.

DESIGN FIRM Pedersen Gesk
ART DIRECTOR Mitch Lindgren
DESIGNER Kris Morgan, Thom Middlebrook
CLIENT Arctco Industries
PRODUCT Arctic Cat

The brand mark created clearly captures the essence of the brand and the attitude of the target consumer. The black package demonstrates the premium, high-tech quality of the products.

DESIGN FIRM Mires Design
ART DIRECTOR José A. Serrano
DESIGNER José A. Serrano, Miguel Perez
PHOTOGRAPHER Carl Vanderschuit
CLIENT Voit Sports
PRODUCT Basketball

Voit Sports was introducing a brand-new line of ball with a unique grip. Mires Design selected a bold, highly visible, and somewhat playful looking font. It then manipulated the original font to emphasize the energy of a three-dimensional sphere-shaped product.

DESIGN FIRM Mires Design
ART DIRECTOR José A. Serrano
DESIGNER José A. Serrano, Miguel Perez
ILLUSTRATOR Tracy Sabin
CLIENT Found Stuff Paper Works
PRODUCT Stationery

The idea behind this project was to develop packaging that accomplished two purposes: (1) to convey the message that the product was made of 100 percent recycled materials, and (2) to create a sense of value for a product that was made of recycled products.

consumer products | 29

DESIGN FIRM Hornall Anderson
　　　　　　　Design Works
ART DIRECTOR Jack Anderson
DESIGNER Jack Anderson, Heidi Favour,
　　　　　　John Anicker
CLIENT OXO International
PRODUCT OXO Good Grips Barbecue tools

A natural Kraft box helped the product stand out among its competitors, which use a more slick and glossed box surface. Although rough in appearance, the closed box treatment on Kraft paper emphasizes a more upscale quality in simplicity, as well as a higher grafe of product quality.

DESIGN FIRM Mires Design
ART DIRECTOR José A. Serrano
DESIGNER José A. Serrano, Miguel Perez
ILLUSTRATOR Tracy Sabin
CLIENT Found Stuff Paper Works
PRODUCT Notepads on 100 percent recycled paper

All of the materials used in both this product and package are recycled and natural. The cotton was organically grown, and the material left over from the cotton was recycled into paper pulp that was later used to make the paper for the sketch books. The labels were printed with soy ink.

DESIGN FIRM Mires Design
ART DIRECTOR José A. Serrano
DESIGNER José A. Serrano
PHOTOGRAPHER Carl Vanderschuit
CLIENT Agassi Enterprises
PRODUCT Fusion gripping powder

This project introduced a new gripping powder with a visual graphic that conveyed a sense of strength. The tight fist and the strong use of bold, simple color achieve this effect.

DESIGN FIRM
Steve Trapero Design
ART DIRECTOR/DESIGNER
Steve Trapero
CLIENT
Western Water International, Inc.
PRODUCT
FrigiPure refrigerator air purifier
TECHNIQUE
Offset

The label and header card were created using QuarkXPress and Adobe Illustrator and Photoshop. The client wanted a design that communicated clean air in a soft way.

consumer products | 31

DESIGN FIRM
CommuniQué Marketing
ALL DESIGN
C. Benjamin Dacus
CLIENT
Carpenter Company
PRODUCT
Foam mattress pad
TECHNIQUE
Offset

This package insert was part of a program to promote products designed for active people. The product also protects against germs and odor-causing bacteria.

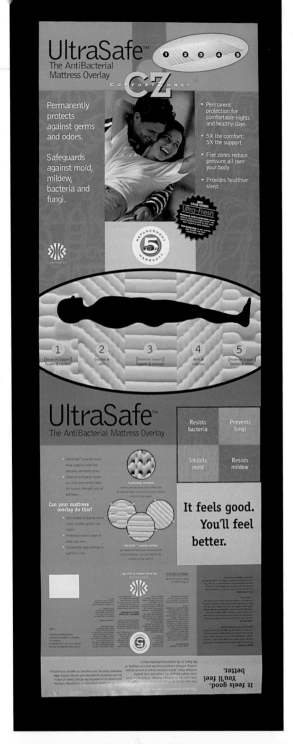

DESIGN FIRM
CommuniQué Marketing
DESIGNER/ILLUSTRATOR
Anne R. Powell
CLIENT
Carpenter Company
PRODUCT
Outback Blues couch cushion
TECHNIQUE
Flexography

Outback Blues is a line of pillows which are faux fleece on one side and faux denim on the other. The illustrations were done by hand and scanned into a document. The final art was created using Adobe Illustrator and QuarkXPress.

DESIGN FIRM After Hours Creative
ART DIRECTOR/DESIGNER After Hours Creative
ILLUSTRATOR Bob Case
CLIENT Kanga Company
PRODUCT Roof Top travel cover
TECHNIQUE Offset

Designed to appeal to the outdoorsy crowd, the Roof Top packaging uses a friendly, wood-cut style to reinforce the product's simplicity. Showing the product in use makes this unusual product immediately understandable.

DESIGN FIRM Greteman Group
ART DIRECTOR Sonia Greteman
DESIGNER James Strange
ILLUSTRATOR C.B. Mordan, Sonia Greteman
CLIENT Hayes Company
PRODUCT Flower & Herb Garden

An ornate English garden was the inspiration for this natural wood flower box. The engraved original illustration is hand-tinted and depicts the actual flowers in the garden. All art was created in Macromedia FreeHand.

consumer products | 33

DESIGN FIRM Love Packaging Group
ALL DESIGN Tracy Holdeman
CLIENT The Hayes Company
PRODUCT Amaryllis & Paperwhite Bulb Gardens
TECHNIQUE Offset

A folding carton stock "wrap" is used instead of a box, allowing the product's charm to show through. Illustrations were done with markers and colored pencils on plain copier paper. Old-fashioned illustrated stickers were added to the end panels and a raffia-ribbon accent and complimentary gardener's journal complete the package.

DESIGN FIRM Love Packaging Group
ALL DESIGN Tracy Holdeman
CLIENT The Hayes Company
PRODUCT Squirrel feeder
TECHNIQUE Flexography (two colors)

The client wanted a package that was unique, "down-home," and that clearly explained the product. Pencil sketches were enlarged, then scanned and placed in a layout in Macromedia FreeHand. The design required no trapping, only overprinting. The "corrugated corn" is produced in the "scrap" area of the cutting die, so the whole package can be produced in one pass in the manufacturing/printing process.

DESIGN FIRM Sayles Graphic Design
ART DIRECTOR/DESIGNER John Sayles
ILLUSTRATORS John Sayles, Jennifer Elliott
CLIENT American Heart Association
PRODUCT All Kids at Heart
TECHNIQUE Screenprinting, offset, thermography

The designer deliberately chose a "childlike" illustration style for the graphics: hearts and faces.

DESIGN FIRM Rocha & Yamasaki
ART DIRECTOR Mauricio Rocha
DESIGNER Mauricio Rocha, Rosemari Yamasaki
CLIENT Cristallo
PRODUCT Easter egg box
TECHNIQUE Offset

The client needed an inexpensive box that appealed to children as well as adults. The window on the edge of the box reveals the product, which is packed with eye-catching French textured paper. The design was created with CorelDraw. The aim was to create a balance between geometric and organic forms. The final design was printed on just one sheet of paper and sealed with one point of glue.

consumer products | 35

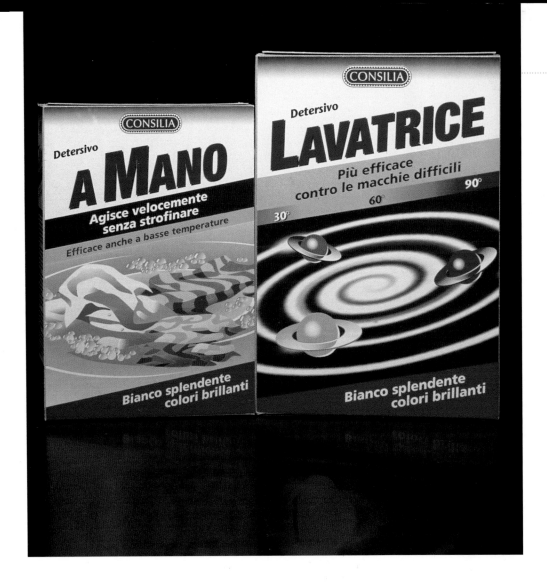

DESIGN FIRM Tangram Strategic Design
ART DIRECTOR/DESIGNER Enrico Sempi
ILLUSTRATOR Guido Rosa, Sergio Quaranta
CLIENT Gruppo SUN
PRODUCT Washing powder
TECHNIQUE Offset

This private label project was started by Michael Peters Group, so it was necessary to follow certain guidelines yet still find a new approach. The product's image and quality is quite high, so Tangram Strategic Design needed to pay special attention to its positioning in the market. Adobe Illustrator and Photoshop were used to create the design.

DESIGN FIRM
Love Packaging Group
ALL DESIGN
Tracy Holdeman
CLIENT
The Hayes Company
PRODUCT
Boot scraper
TECHNIQUE
Offset

Because of budgetary constraints, the designers created packaging that would suit multiple boot scrapers and also function as a point-of-purchase display. A simple litho-label was applied to the front of the countertop POP to speak for all of the products. The illustration was scanned and turned into vectors in Adobe Streamline. The final design was assembled in Macromedia FreeHand.

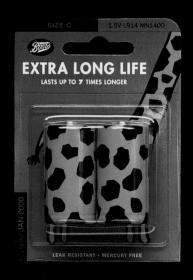
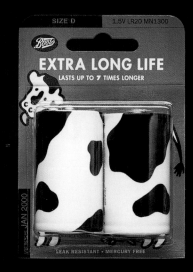
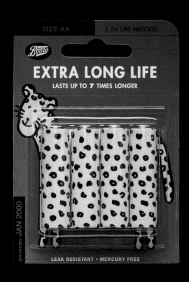

DESIGN FIRM Lewis Moberly
ART DIRECTOR Mary Lewis
DESIGNER/ILLUSTRATOR Shaun Bowen
CLIENT The Boots Company Place
PRODUCT Children's batteries

Boots Batteries is a range of batteries targeted at children. The brand aims to extend gift buying (they are positioned in the store next to children's gifts) and build loyalty to this growing sector. The packaging features different animals whose bodies become the batteries themselves.

DESIGN FIRM
Graef & Ziller Design
ART DIRECTOR
B. Ziller
DESIGNER/ILLUSTRATOR
Andrew Richards
CLIENT
Virtual Motion
PRODUCT
Software
TECHNIQUE
Offset

The box was primarily designed to be sent to Original Equipment Manufacturers (OEMs) with modem hardware. It was also being used for consumer packaging, so the design needed to stand out in a software catalog. The art was created in Adobe Illustrator and printed in five PMS colors on a direct digital Heidelburg press.

consumer products | 37

DESIGN FIRM Hornall Anderson Design Works
ART DIRECTOR Jack Anderson
DESIGNER Jack Anderson, David Bates
CLIENT Smith Sport Optics, Inc.
PRODUCT Sunglasses

The strategy here was to capture the pureness of the outdoors in the merchandising of the company's products and to differentiate the company from its competitors. Natural materials were blended with the actual construction technology and branded with the Smith identity. Old-fashioned metal fasteners and tape are exposed, creating an honest presentation for the packaging.

DESIGN FIRM Sibley-Peteet Design
ART DIRECTOR Jim Bremer, David Beck
DESIGNER/ILLUSTRATOR David Beck
CLIENT Milton Bradley
PRODUCT Game
TECHNIQUE Offset

"Crack the Case" is an adult party game based on different crime solving scenarios. The initial pencil sketches were scanned, and the final art was done in Adobe Illustrator.

38 | Package & Label Design

DESIGN FIRM
Planet Design Company
ART DIRECTOR
Dana Lytle, Kevin Wade
DESIGNER
Dana Lytle, Raelene Mercer
CLIENT
Gräber USA
PRODUCT
Outback bike racks
TECHNIQUE
Lithowrap, offset; box, flexography

Planet Design Company's challenge was creating packaging that evoked all the outdoor imagery associated with biking while fitting into the mass merchandise environment. The design was created with QuarkXPress.

DESIGN FIRM
Sibley-Peteet Design
ART DIRECTOR
Donna Aldridge
DESIGNER
Donna Aldridge, Diane Fannon
ILLUSTRATOR
Donna Aldridge
CLIENT
The Image Bank
PRODUCT
Video
TECHNIQUE
Offset

The client wanted to promote its sports photography, illustration, and film, with a very interactive piece that left the recipient with a fun premium. The package unfolds to reveal a videocassette of sports footage and a pocket folder containing twelve "trading" cards with different sports imagery. The design was created with QuarkXPress and Adobe Illustrator.

consumer products | 39

DESIGN FIRM Animus Comunicação
ART DIRECTOR Rique Nitzsche
DESIGNER Rique Nitzsche, Felicio Torres
CLIENT Suzano
PRODUCT Colored paper
TECHNIQUE Four-color flexography

Because approximately ninety percent of this product's sales is business-to-business and it faces fierce competition, the packaging needed to create brand identity and consumer recognition. Animus Comunicação created a simple display where the product would have a permanent sales place separate from the visual chaos of the store and would be shown in a more apparent and completely new way for 500-sheet packages.

DESIGN FIRM Watts Graphic Design
ALL DESIGN Helen Watts, Peter Watts
CLIENT Intergrain Timber Finishes
PRODUCT Intergrain Stains

The beauty of the timber scene in this simple, clean design helps catch the eye of the customers and create a distinct identity on the shelf.

40 | Package & Label Design

DESIGN FIRM
Martin Ross Design
DESIGNER
Ross Rezac, Martin Skoro
ILLUSTRATOR
Martin Skoro
CLIENT
Cray Research
PRODUCT
Software

The imagery for this package was created from an existing sky photo that Cray Research had used for other materials. Martin Ross Design created the globe graphic and incorporated it into the sky photo by using Adobe Photoshop.

DESIGN FIRM Martin Ross Design
DESIGNER Ross Rezac, Martin Skoro
ILLUSTRATOR Martin Skoro
CLIENT St. Paul Companies
PRODUCT Disaster Recovery package

Martin Ross Design created the architecture of this package in order to make disks, forms, and information easily accessible. The imagery was done in Adobe Photoshop using photos taken especially for the piece.

consumer products | 41

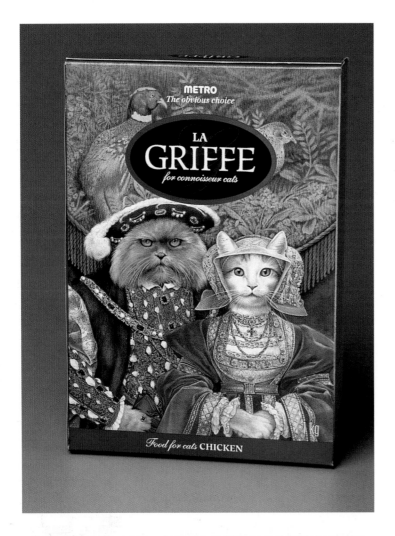

DESIGN FIRM Goodhue & Associés
ART DIRECTOR Suzanne Côté
DESIGNER Suzanne Côté, Nicole Bouchard
ILLUSTRATOR Suzanne Duranceau
CLIENT Épiciers unis Métro-Richelieu
PRODUCT Cat food

This design concept was inspired by the paintings of Magritte, Gainsborough, and Holbein. The name La Griffe highlights the aristocratic character of cats. The illustration contains graphic references to the type of food.

DESIGN FIRM Goodhue & Associés
ART DIRECTOR Suzanne Côté
DESIGNER Suzanne Côté, Andre Saint-Loup
ILLUSTRATOR Louis Prud'homme
CLIENT Épiciers unis Métro-Richelieu
PRODUCT Coffee filters

Goodhue & Associés used an original and interesting approach to photographing the product. Pastel colors ensure that it stands apart from the competition.

consumer products | 43

Hairworks SHAMPOO

Normal Hair

Works With Nature
To Fortify

Health & Fashion

DESIGN FIRM FRCH Design Worldwide
ART DIRECTOR Joan Donnelly
DESIGNER Tim A. Frame
PHOTOGRAPHER Bray Ficken
CLIENT Aca Joe
PRODUCT Men's jeans
TECHNIQUE Four-color offset

These pocket flashers are used to communicate four types of jean fits for men's denim pants. Different images of women's lips were used to represent individual fits.

DESIGN FIRM FRCH Design Worldwide
ART DIRECTOR Joan Donnelly
DESIGNER Tim A. Frame
CLIENT Aca Joe
PRODUCT Men's chinos
TECHNIQUE Two-color offset
(one color plus timed varnish)

These waistband tickets are used to communicate two styles of men's basic chinos.

DESIGN FIRM Haley Johnson Design Company
DESIGNER/ILLUSTRATOR Haley Johnson
CLIENT Bath & Body Works
PRODUCT Body cream
TECHNIQUE Flexography

These small size body cream tubes were introduced to create excitement for an existing product line. Each flavor displays appropriate iconography in a quilt-like pattern. Hand-written typography is used to emphasize this hand-touched approach. Final art was created in Adobe Illustrator.

DESIGN FIRM Haley Johnson Design Company
DESIGNER/ILLUSTRATOR Haley Johnson
CLIENT Makeup Group
PRODUCT Stila loose powder
TECHNIQUE Offset

This utilitarian, recyclable package for Stila loose powder aims to differentiate itself from traditional over-packaged cosmetic products. Final art was created in Adobe Illustrator. This is a high-end product.

health & fashion | 47

DESIGN FIRM Shimokochi/Reeves
ART DIRECTOR Mamoru Shimokochi, Anne Reeves
DESIGNER Mamoru Shimokochi
ILLUSTRATOR Jim Krogle
CLIENT Leiner Health Products
PRODUCT Bodycology
TECHNIQUE Lithography

This is the brand identity and package redesign for thirty-three body, bath, and haircare products. To achieve maximum shelf impact, Shimokochi/Reeves used large, brightly-colored illustrations and strengthened the Bodycology brand logotype. The frosted containers and floral images reflect each product's natural ingredients and fragrance. The art was created with Adobe Illustrator.

DESIGN FIRM Shimokochi/Reeves
CREATIVE DIRECTOR Mamoru Shimokochi, Anne Reeves
DESIGNER/ILLUSTRATOR Mamoru Shimokoch
CLIENT Dep Corporation
PRODUCT Nature's Family

This is a proposed brand identity and package revitalization for a line of natural ingredient–based skin care products. The art was created with Adobe Illustrator.

DESIGN FIRM Jeff Labbé Design Company/dGWB Advertising
ART DIRECTOR/DESIGNER Jeff Labbé
CLIENT Shimano Bicycling
PRODUCT Shimano shoes and components
TECHNIQUE Offset

The designers developed the tags utilizing two PMS colors to save costs. By reversing out black halftones and trapping yellow ones, it was able to give the design a clean but urban aggressive feel. The laminated tags can be easily marked and erased whenever pricing changes for sales and promotions.

DESIGN FIRM Jeff Labbé Design Company/dGWB Advertising
ART DIRECTOR Jeff Labbé, Wade Kaniakowsky
DESIGNER/ILLUSTRATOR Jeff Labbé
CLIENT Vans Shoes Inc.
PRODUCT Vans sandals
TECHNIQUE Flexography, screen

The box took on an attitude of its own with bulky graphics, fun, uninhibited, seasonal photography, colors that reflect the earth tone palette of the sandals, and a theme of casualness. For the printing, the designer used an existing box die and crashed it two-color on Kraft paper allowing for fun imperfections to appear.

DESIGN FIRM Lewis Moberly
ART DIRECTOR Mary Lewis
DESIGNER Nun Glaister
CLIENT Alfred Dunhill, Ltd.
PRODUCT Dunhill "d"

The sharp champagne-silver box projects a strong, modern image that reflects the character of the scent. The matching carton and bottle were designed to appeal to a broad base of people.

DESIGN FIRM Able Design
ART DIRECTOR/DESIGNER Martha Davis
CLIENT Hanna Company
PRODUCT Epilogue skincare
TECHNIQUE Screen

The project objectives were: (1) to communicate that egg yolk extract is the basis of the products, (2) to emphasize the purity of the natural ingredients and processing, and (3) to create a premium package suitable for both the Asian and European/U.S. markets. Software used included Adobe Illustrator and Photoshop and Vellum.

DESIGN FIRM Curtis Design
ART DIRECTOR David Curtis
DESIGNER Matt Sullivan
CLIENT Weider Nutrition
PRODUCT Prime Time nutrition program

Prime Time was developed as a high-end nutritional program exclusively for men. Actor Robert Urich endorses the product. Premium printing techniques were used: eight colors, embossing, and foilized type.

DESIGN FIRM Hans Flink Design Inc.
ART DIRECTOR Hans D. Flink
DESIGNERS Chang-Mei Liu, Mark Krukonis
CLIENT Chesebrough-Ponds
PRODUCT Pond's cosmetic line

Global brand identity development and graphic system design for Pond's new AHA products line helped the brand to dramatically regain market share. The art was created on a Macintosh using Adobe Freehand.

DESIGN FIRM Louisa Sugar Design
ART DIRECTOR/DESIGNER Louisa Sugar
ILLUSTRATOR Susan Bercu
CLIENT Golden Neo-Life Diamite International
PRODUCT GNLD nutritional supplements

The company's orange and green logo colors are used on both labels to reinforce a new global brand identity. For the vitamin line, orange predominates, with red accents and a brush illustration of fruit to reflect the healthy, whole ingredients inside. The deep green herbal complex label carries over the premium look, with gold accents to indicate that this is a special kind of supplement.

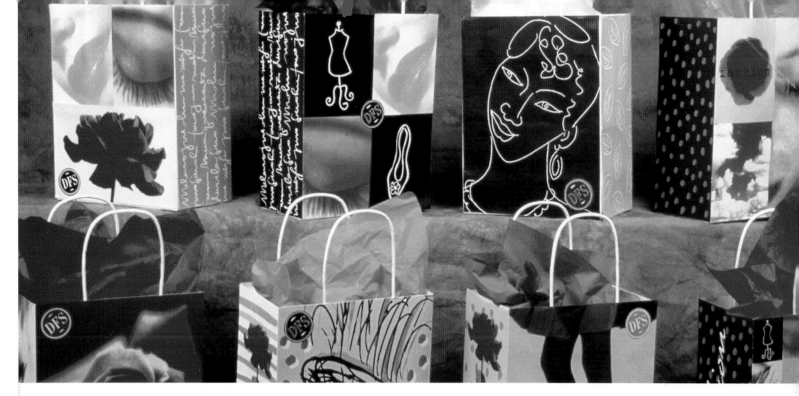

DESIGN FIRM Morla Design
ART DIRECTOR Jennifer Morla
DESIGNER Jennifer Morla, Craig Bailey, Petra Geiger
ILLUSTRATOR Craig Bailey, Petra Geiger, Matisse
CLIENT DFS Group Limited
PRODUCT Women's shopping bags for duty free shops
TECHNIQUE Offset

These bright, eye-catching, and distinctive shopping bags were created for the duty free women shoppers. Images allude to the body, to movement, and to a general sense of playfulness.

DESIGN FIRM
Pentagram Design Inc.
ART DIRECTOR
Paula Scher
DESIGNER
Lisa Mazur
CLIENT
G.H. Bass & Co.
PRODUCT
Shoes and apparel
TECHNIQUE
Offset

A refined graphic approach signifies Bass's new direction in retailing shoes and apparel. Pentagram reinterpreted an old, undisciplined collection of graphics, advertising, and packaging into a coordinated system of imagery that celebrates Bass's strong New England heritage.

health & fashion | 53

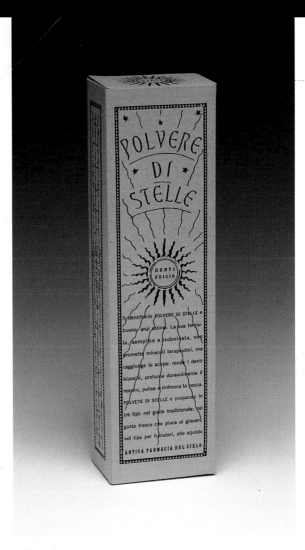

DESIGN FIRM Louise Fili Ltd.
ART DIRECTOR/DESIGNER Louise Fili
CLIENT El Paso Chile Company
PRODUCT Polvere di Stelle toothpaste
TECHNIQUE Offset

The client wanted a package design to evoke turn-of-the-century Italy. Louise Fili Ltd. supplied the name, Polvere di Stelle (stardust), and designed the package after researching the design period.

DESIGN FIRM Rickabaugh Graphics
ART DIRECTOR/DESIGNER Eric Rickabaugh
CLIENT The Limited Too
PRODUCT Work Wear Jeans
TECHNIQUE Embroidered patches

This line of jeans for young girls was built around the concept of "painter" jeans with multicolored patches. The embroidered patches were designed to capture a "funky" retro look and were created in Macromedia FreeHand.

DESIGN FIRM Robilant & Associati
ART DIRECTOR Maurizio di Robilant
DESIGNER Lucia Sommaruga
CLIENT Koh-I-Noor Italy
PRODUCT Wood collection

The project's aim was to create natural, environmentally-friendly packaging for a line of combs and brushes made of natural materials (wood, hair). Instead of using acetate, the designers used drawings to represent the product contained in the box. The whole package needs only one glue point.

DESIGN FIRM Teikna
ART DIRECTOR/DESIGNER Claudia Neri
CLIENT Club Monaco
PRODUCT Jeans Tag
TECHNIQUE Offset

For their Fall '96 jeans, Club Monaco needed a design for their pant tags which would fit with the retailer's simple, modern image, and appeal to men. Artwork was done in QuarkXPress.

health & fashion | 55

DESIGN FIRM Angelo Sganzerla
ART DIRECTOR/DESIGNER Angelo Sganzerla
ILLUSTRATOR Alfonso Goi
CLIENT L'Erbolario
PRODUCT Toiletries

The watercolor illustration for this sandalwood bath product line shows a typical native landscape.

DESIGN FIRM
Angelo Sganzerla
ART DIRECTOR/DESIGNER
Angelo Sganzerla
ILLUSTRATOR
Gennadij Spirin
CLIENT
L'Erbolario
PRODUCT
Perfume and toiletries

The illustration for this L'Erbolario Christmas gift tin is a watercolor depicting fairy-tale scenes.

DESIGN FIRM
DesignCentre of Cincinnati
ART DIRECTOR
Kamren Colson
DESIGNER
Joan Bishop Gottsacker
ILLUSTRATOR
Donna Talerico
CLIENT
Nine West Group Inc.
PRODUCT
Easy Spirit Once-a-Year Sale bag
TECHNIQUE
Offset

For this annual sales event, Nine West gave DesignCentre the freedom to create a graphic look that would build on the new brand personality of Easy Spirit. It chose a contemporary, fun, and colorful approach for this promotion, providing an irresistible invitation for both new and existing customers to come into the store.

DESIGN FIRM
Vanessa Eckstein
ALL DESIGN
Vanessa Eckstein
CLIENT
Natura
PRODUCT
Beauty product
TECHNIQUE
Offset

Aimed at a younger, more environmentally conscious group of people, this product tries to express the importance of environmental issues and the company's respect for the environment without overdoing the "recycled" look. It's an energetic approach at a real philosophy.

health & fashion | 57

DESIGN FIRM
Vanessa Eckstein
ALL DESIGN
Vanessa Eckstein
CLIENT
The Bag Stand Company
PRODUCT
Cosmetic cases
TECHNIQUE
Offset

This product is aimed at a teenage market. The design had to be playful, hip, and colorful. It's a "fun" product, and the design needed to express the energetic feeling of this young audience.

DESIGN FIRM Mires Design
DESIGNER José A. Serrano, Miguel Perez
ILLUSTRATOR Tracy Sabin
CLIENT Bordeaux Printers
PRODUCT Bordeaux printing services

This client wanted to communicate the image of high-end printing. Mires Design did this by printing labels and hang tags as part of a quality assurance program. The labels and tags were signed by sales representatives and production people to reassure clients that the printed samples had been carefully inspected.

DESIGN FIRM
revoLUZion
ALL DESIGN
Bernd Luz
CLIENT
World Merchandising Corporation
PRODUCT
Clothes and bags in hemp
TECHNIQUE
Watercolor, offset

The Hanf (hemp) type and logo on labels and buttons were developed for three collections of clothes and bags, all made from hemp. Macromedia FreeHand was used to create the design.

DESIGN FIRM Rocha & Yamasaki
ART DIRECTOR Mauricio Rocha
DESIGNER Mauricio Rocha, Rosemari Yamasaki
CLIENT Punto Smock
PRODUCT Packing box

The ornaments are based on punto smock, a kind of stitch applied on the clothes. The squares represent a kind of tissue pattern often used with punto smock. The client needed an inexpensive, one-color package. The solution was to use halftone color to give more contrast and movement to the box. CorelDraw was used to create the design.

health & fashion | 59

DESIGN FIRM The Design Company
ART DIRECTOR Marcia Romanuck
DESIGNER/ILLUSTRATOR Alison Scheel
CLIENT Vitamin Healthcenters
PRODUCT Herbal formulas
TECHNIQUE Offset

These labels were designed to work with the Vitamin Healthcenter's identity in color and feel, yet look distinct in content. The various herbs were created in Adobe Illustrator and then manipulated in QuarkXPress. There are about twenty products in this line.

DESIGN FIRM Malik Design
ALL DESIGN Donna Malik
CLIENT Franco Manufacturing
PRODUCT Kitchen linens
TECHNIQUE Offset

Adobe Photoshop and Macromedia FreeHand were used to create a distinctive hang tag/header card for a new line of kitchen linens (towels, pot holders, placemats, curtains, and so on).

60 | Package & Label Design

DESIGN FIRM
Greteman Group
ART DIRECTOR
Sonia Greteman
DESIGNER
Sonia Greteman
CLIENT
Chica Bella
PRODUCT
Natural beauty products

This line of Chica Bella natural beauty products is sold in health food stores and was designed to echo the colors and patterns of nature. Each product line encompasses an appropriate feeling of imagery to the botanical composition. It was created in Macromedia FreeHand.

DESIGN FIRM Malik Design
ALL DESIGN Donna Malik
CLIENT Franco Manufacturing
PRODUCT Kitchen linens
TECHNIQUE Offset

Adobe Photoshop and Macromedia FreeHand were used to create a distinctive hang tag/header card for a new line of kitchen linens (towels, pot holders, placemats, curtains, and so on).

health & fashion | 61

DESIGN FIRM
Desgrippes Gobé & Associates
CREATIVE DIRECTOR
Peter Levine
ART DIRECTOR
Sarah Allen
DESIGNER
Michael Milley
CLIENT
FILASport
PRODUCT
Sportswear

The client wanted a brand positioning that combined the attributes of European athletic wear with American men's sportswear markets. The challenge was to establish a visual territory and graphics program which communicated the company's spirit: energy, dynamism, and positivity. The FILASport garment logos were designed to lend authenticity and credibility to the individual garments.

DESIGN FIRM
Desgrippes Gobé & Associates
DESIGN DIRECTOR
Susan Berson
CLIENT
The Limited, Inc.
PRODUCT
Bath & Body Works Hairworks

Desgrippes Gobé & Associates combined the clean, white packaging of traditional performance haircare products and the hand-touched quality of Bath & Body Works' "from the Heartland" positioning to develop a distinctive packaging that introduced the performance aesthetic to Bath & Body Works.

DESIGN FIRM Desgrippes Gobé & Associates
CREATIVE DIRECTOR Peter Levine, Kenneth Hirst
DESIGN DIRECTOR Frances Ullenberg
DESIGNER Christopher Freas
CLIENT Ann Taylor Inc.
PRODUCT Ann Taylor Destination fragrance

This packaging reinforces the Ann Taylor brand identity by communicating a natural, honest design that is consistent with the store's total retail identity. The fragrance packaging's materials are made from recycled and recyclable materials and have soft, nature-based shapes.

DESIGN FIRM Sibley-Peteet Design
ART DIRECTOR Bryan Jesse, Rex Peteet
DESIGNER Rex Peteet, Derek Welch
ILLUSTRATOR Stephen Alcorn
CLIENT Farah
PRODUCT Men's clothing
TECHNIQUE Offset, weaving

Sibley-Peteet Design created a new identity for a younger market. The client wanted a solid, bold, simple feel. Woodcuts were scanned into Adobe Illustrator.

health & fashion | 63

DESIGN FIRM Sibley-Peteet Design
ART DIRECTOR Bryan Jesse
DESIGNER/ILLUSTRATOR Rex Peteet
CLIENT Farah
PRODUCT Men's clothing
TECHNIQUE Offset, weaving

Sibley-Peteet Design embraced the existing equity of the typeface and updated the look with a contemporary and more fashionable treatment without turning off the older, existing market. The art was created with Adobe Illustrator.

DESIGN FIRM Sibley-Peteet Design
ART DIRECTOR Bryan Jesse, Rex Peteet
DESIGNER/ILLUSTRATOR Rex Peteet, Derek Welch
CLIENT Farah
PRODUCT Men's clothing
TECHNIQUE Offset, weaving

Sibley-Peteet Design created a new identity and image for a young, mobile market. The workplace is becoming more casual and Savane meets the need for fashion that is comfortable, good looking, and easy to care for. Many products in this line are wrinkle-free and stain resistant.

DESIGN FIRM Planet Design Company
ART DIRECTOR Dana Lytle, Kevin Wade
DESIGNER Kevin Wade, Martha Graettinger
CLIENT Lortex
PRODUCT Natural nylon fabric
TECHNIQUE Screen

This piece, used for presentation by the sales staff, tries to capture both the high tech and natural characteristics of the nylon. Macromedia FreeHand and QuarkXPress were used to create this design.

health & fashion | 65

Food

DESIGN FIRM Boulder Design & Illustration
ALL DESIGN Kristin Geishecker
CLIENT Community Products, Inc.
PRODUCT Rainforest Crunch
TECHNIQUE Offset

The client needed a bright, vivid rainforest design that would jump off shelves. The packaging design was created by using Adobe Illustrator and Photoshop, and the illustrations were done in watercolor. The final pieces were very well received.

DESIGN FIRM Boulder Design & Illustration
ALL DESIGN Kristin Geishecker
CLIENT Community Products, Inc.
PRODUCT Sweet River Chocolates
TECHNIQUE Offset

Sweet River Chocolate boxes and labels were created using Aldus Freehand. Illustrations were done by hand and in Aldus Freehand. The final packages were very well received.

DESIGN FIRM Damion Hickman Design
ART DIRECTOR/DESIGNER Damion Hickman
CLIENT Napa Valley Gourmet Salsa
PRODUCT Gourmet salsa
TECHNIQUE Letterpress

This client needed a label to convey that the product came from Napa Valley. Because the product contains wine in the ingredients, the concept of a woodcut looking label (similar to the signs found at a winery entrance) was used to set the product apart on the shelf from normal "hot and spicy" salsas.

DESIGN FIRM Minkus & Associates
ART DIRECTOR Robert Minkus
DESIGNER Deborah McSorley
CLIENT Gap International
PRODUCT Power Pasta

Power Pasta is a high-protein pasta targeted to health-conscious individuals who shop at specialty and health food stores. The design was created by using Adobe Illustrator 6.0, and the label's vibrant colors jump out from the earth tones that dominate the health food category.

food | 69

DESIGN FIRM Primo Angeli Inc.
ART DIRECTOR Primo Angeli, Carlo Pagoda
DESIGNER Philippe Becker
CLIENT Harden & Huyse
PRODUCT Chocolates

This design establishes an exclusive, high-end image for the brand on the same level with top international chocolatiers. The design's subtle, melting chocolate-colored swirls communicate a mouth-watering, but refined, taste appeal highlighted by the brand name in elegant gold lettering. Modular containers were designed for both display and cost-effective shipping.

DESIGN FIRM Haley Johnson Design Company
DESIGNER/ILLUSTRATOR Haley Johnson
CLIENT Amazing Grazing
PRODUCT Gourmet Snackin' Sauce
TECHNIQUE Flexography

This package focuses on the blending of international flavors used to create the sauce. Icons from around the world are used to suggest an international experience of snacking around the globe. The colorful labels are accompanied by custom printed lids.

DESIGN FIRM Watts Graphic Design
ART DIRECTOR/DESIGNER Helen and Peter Watts
CLIENT Pond Cottage
PRODUCT Pond Cottage eggs
TECHNIQUE Offset

Guests of this bed and breakfast inn are treated to fresh eggs every morning wrapped with this attractive label printed on specialty paper. It has been very well received and successful.

DESIGN FIRM Watts Graphic Design
ART DIRECTOR/DESIGNER Helen and Peter Watts
CLIENT Pond Cottage
PRODUCT Pond Cottage jam
TECHNIQUE Offset

This boutique label for a unique country bed and breakfast retreat gave added appeal to its decor. This product is given as a gift to the people who visit. Extra packs are available for a small fee. It was a very successful promotion.

food | 71

DESIGN FIRM Shimokochi/Reeves
ART DIRECTOR Mamoru Shimokochi, Anne Reeves
DESIGNER Mamoru Shimokochi
CLIENT Camino Real Foods, Inc.
PRODUCT Las Campanas taquitos
TECHNIQUE Flexography

The design of this new packaging for a line of high-quality taquitos was created in Adobe Illustrator. The packaging is part of a program involving more than fifty products.

DESIGN FIRM
Del Rio Diseño
ART DIRECTOR/DESIGNER
Del Rio Diseño team
ILLUSTRATOR
Carlos Rojas
CLIENT
Maggi
PRODUCT
Maggi soup
TECHNIQUE
Rotogravure

The emphasis here is on the Maggi brand, associating the origin of each variety of soup with a place in the world. An appetizing presentation of the product, which highlights its main ingredient, supports the product's image.

ART DIRECTOR David Curtis
DESIGNER/ILLUSTRATOR Chris Benitez
CLIENT Nibbler Farms
PRODUCT Fresh cut produce

packaging relevance in today's fresh cut produce category.

DESIGN FIRM
Del Rio Diseño
ART DIRECTOR/DESIGNER
Del Rio Diseño team
CLIENT
Carozzi
PRODUCT
San Remo pastas
TECHNIQUE
Rotogravure

This line has been redesigned to make the brand remind the consumer of Italy and its tradition in pastas. Strong diagonals and the colors of the Italian flag were used in the design to distinguish it from the competition.

food | 73

DESIGN FIRM Lambert Design
ALL DESIGN Christine Lambert
CLIENT Duo Delights
PRODUCT White chocolate bark
TECHNIQUE Offset printing

The challenge for the designer was to create a package that conveys the soft, fruity character of the product in a form that would be inexpensive to produce and easy for the client to assemble. These small, colorful pieces definitely stand out in the clutter of the specialty foods market. All art and type was created in Illustrator.

DESIGN FIRM The Coleman Group
ART DIRECTOR Edward Morrill
DESIGNER William Lee
ILLUSTRATOR J.C. Chou
CLIENT Biscotti & Co.
PRODUCT Karen's Biscotti
TECHNIQUE Offset

This package structure had to be cost-effective in terms of both materials and filling, be reclosable, provide a large graphic area, and stand easily on retail shelves. The casual contemporary graphics are reflective of coffee bar decor with a light and airy, on-the-go, appetizing appearance.

DESIGN FIRM
Paul Ng Design & Productions
ART DIRECTOR/DESIGNER
Paul Ng
CLIENT
Sun Ming Hong (Canada) Ltd.
PRODUCT
Canadian Ginseng Candy

The history of how ginseng was founded and developed by the health food industry is shown on the box cover. Gold foil full-color printing make a solid impact in the market.

food | 75

DESIGN FIRM Louise Fili Ltd.
ART DIRECTOR/DESIGNER Louise Fili
CLIENT Grafton Goodjam
PRODUCT Vinegar
TECHNIQUE Offset

This was a new lower-priced line of gourmet vinegars produced by Grafton Goodjam. An old photograph was used to give the company a more personal identity.

DESIGN FIRM Louise Fili Ltd.
ART DIRECTOR/DESIGNER Louise Fili
CLIENT Picholine
PRODUCT Olives
TECHNIQUE Offset

Picholine olives are sold in a restaurant of the same name in New York City. The label features the restaurant logo and graphics.

DESIGN FIRM CMA
ART DIRECTOR/DESIGNER Bob Milz
ILLUSTRATOR Jeff Sanson, Steve Wells
CLIENT Sysco
PRODUCT Sysco flavored vinegars
TECHNIQUE Offset, watercolor illustration

This package was designed for Sysco's customers as well as to be displayed in front of the restaurant or behind counters. Sysco asked CMA to create a format with an old-world look that could be extended as more flavors were added.

DESIGN FIRM Louise Fili Ltd.
ART DIRECTOR/DESIGNER Louise Fili
CLIENT Beantown Soup Co.
PRODUCT Beantown Soup labels
TECHNIQUE Offset

The client needed an all-purpose label to be used for seeds, soup mixes, cornbread mixes, and so on. Because the company has a large mail-order audience, a commercial look was unnecessary.

DESIGN FIRM CMA
ALL DESIGN Bob Milz
CLIENT Frank & Bryan's Foods
PRODUCT Frank & Bryan's SunSalsa
TECHNIQUE Illustration in Macromedia FreeHand, offset

The maker of this salsa requested a hand-crafted, Southwestern style treatment for its labels. After evaluating several traditional sun images, CMA settled on a simple still life. The range of flavors are distinguished by color bands around the trademark.

DESIGN FIRM CMA
ART DIRECTOR Bob Milz
DESIGNER Dean Narahara
PHOTOGRAPHER David Stoecklein
CLIENT Double B Foods
PRODUCT Texas Pride beef jerky
TECHNIQUE Flexography

Double B Foods sells a lot of high-quality, natural-style beef jerky through retailers based in Japanese airports. Because of the popularity of authentic Texas products in this venue, the client requested that unmistakable Texas iconography play a major roll in the design.

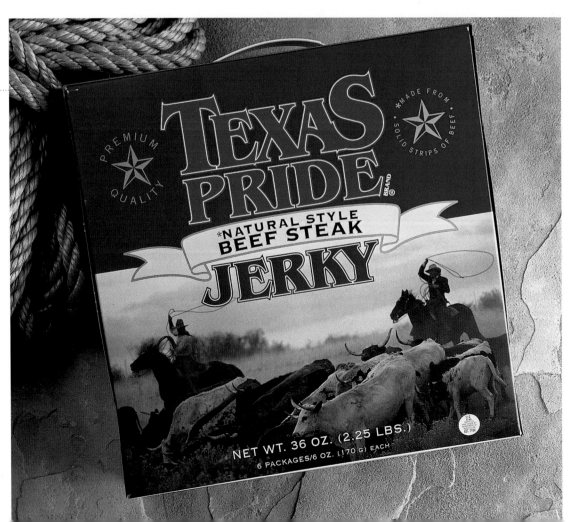

78 | Package & Label Design

DESIGN FIRM CMA
ART DIRECTOR Bob Milz
DESIGNER Dave Campbell, Bob Milz
ILLUSTRATOR Bob Milz, Dave Campbell, Trish Hill
CLIENT Uncle Ben's Rice
PRODUCT Uncle Ben's Beans & Rice line
TECHNIQUE Macromedia FreeHand

For a new line of regional specialty beans and rice dishes, Uncle Ben's asked CMA to combine its corporate orange box look with a decorative illustrative element to romanticize the regional differences among the various beans and rice combinations.

DESIGN FIRM The Hive Design Studio
DESIGNER Laurie Okamura, Amy Bednarek
CLIENT K.J.B. Foods
PRODUCT Old Wharf Fish House brand specialty foods
TECHNIQUE Flexography

The "Old World Fishing Harbor" look in a "fishing crate" package was created by using Adobe Illustrator with a rubber stamped look. Burnt paper was scanned to create the old paper look on the labels.

food | 79

DESIGN FIRM Carmichael Lynch
ALL DESIGN Peter Winecke
CLIENT Cargill Foods
PRODUCT Flour
TECHNIQUE Web

Progressive Baker is a new line of flour for commercial use only. The design had to be solid as well as eye-catching.

DESIGN FIRM Angelo Sganzerla
ART DIRECTOR/DESIGNER Angelo Sganzerla
ILLUSTRATOR Franco Testa
CLIENT Stainer
PRODUCT Paté, salami, sauces, fruit

A black line drawing was used for this line of hand-crafted specialties. The different colors represent the various lines.

DESIGN FIRM Angelo Sganzerla
ART DIRECTOR/DESIGNER Angelo Sganzerla
ILLUSTRATOR Roberto Weikmann
CLIENT Stainer
PRODUCT Pralines, meringues, nougats

The illustrations for this children's package line are collages showing fantasy patterns or scenes evocative of Mexico, the land of cocoa's origin. The rear illustration shows the product, which is always cocoa-based.

DESIGN FIRM Angelo Sganzerla
ART DIRECTOR/DESIGNER Angelo Sganzerla
ILLUSTRATOR Alfonso Goi
CLIENT Stainer
PRODUCT Praline chocolates

A line of praline chocolates made with natural ingredients, such as orchard fruits and herbs, are illustrated with watercolors.

food | 81

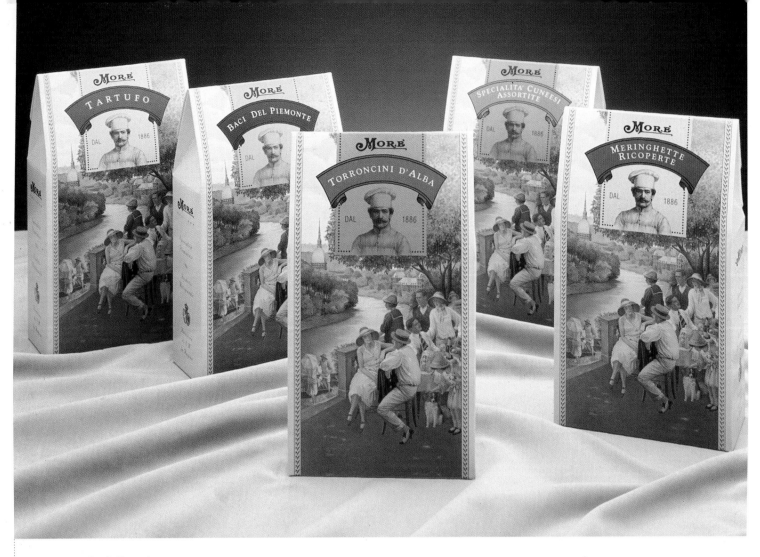

DESIGN FIRM Angelo Sganzerla
ART DIRECTOR/DESIGNER Angelo Sganzerla
ILLUSTRATOR Alfonso Goi
CLIENT Stainer
PRODUCT Pralines, meringues, nougats

The watercolor illustration for this line of typical Piedmontese sweets depicts a riverside festivity against the background of Turin.

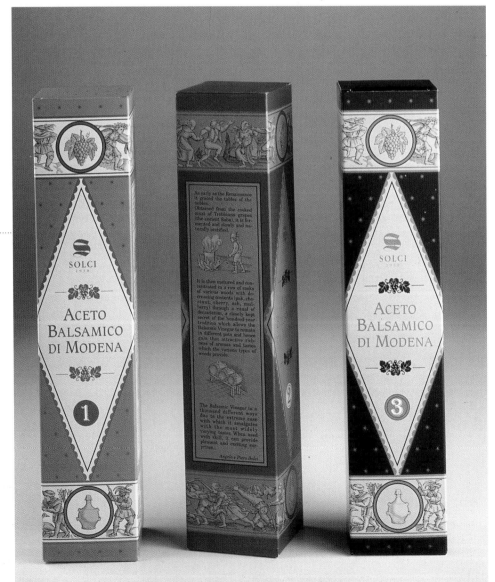

DESIGN FIRM Angelo Sganzerla
ART DIRECTOR/DESIGNER Angelo Sganzerla
ILLUSTRATOR Alfonso Goi
CLIENT Solci
PRODUCT Modena Balsamic Vinegar

Here are three gift packs of Modena Balsamic Vinegar, the finest produced in Italy. The various colors on the box correspond with the different prices according to length of aging.

DESIGN FIRM Maddock Douglas, Inc.
ALL DESIGN Maddock Douglas, Inc.
CLIENT Select Brands
PRODUCT Dip-N-Bread
TECHNIQUE Offset, four-color process

This design was created to appeal to the broad-based consumer market of wholesale clubs. Attention to color was given to create impact in the freezer case as well as an Italian feel.

DESIGN FIRM
Maddock Douglas, Inc.
ART DIRECTOR
Maddock Douglas
DESIGNER
Maddock Douglas
CLIENT
Michael Season Ent.
PRODUCT
Shape-Ups
TECHNIQUE
Flexography

Maddock Douglas named and designed the Shape-Ups packages to be fun, colorful, and vibrant in an otherwise drab product category.

DESIGN FIRM Pedersen Gesk
ART DIRECTOR Rony Zibara
DESIGNER/ILLUSTRATOR Roger Remaley
CLIENT Cargill Foods
PRODUCT Balboa's Sea Salt

Pedersen Gesk positioned this product as the exciting alternative to a commodity item by creating a distinctive name, brand mark, and color palette that accentuated the sea salt's San Francisco heritage.

DESIGN FIRM Ad One
ART DIRECTOR Tom Stoerrle
DESIGNER/ILLUSTRATOR Susan Healy
CLIENT Terrace Il Ristorante, Century Plaza Hotel & Tower
PRODUCT Olive and garlic relish
TECHNIQUE Offset lithography

Olivo Celestiale was created to provide the hotel's director of operations with a special food and beverage gift to present to VIP guests, to market as a "take-out item," and to promote the "cucina rustica" of Terrace Il Ristorante.

DESIGN FIRM Pedersen Gesk
ART DIRECTOR Rony Zibara
DESIGNER Kris Morgan
CLIENT Schwan's Sales Enterprises
PRODUCT Freschetta pizza

To introduce Schwan's new frozen bake-and-rise pizza, Pedersen Gesk created a brand identity that communicates the product's fresh flavor. The logo is reminiscent of a flour mark to represent baking, while the cutaway reveals appetizing toppings and a thick crust.

food | 85

DESIGN FIRM Rocha & Yamasaki
ART DIRECTOR Mauricio Rocha
DESIGNER Mauricio Rocha, Rosemari Yamasaki
CLIENT Cristallo
PRODUCT Petit-Four can
TECHNIQUE Offset with Pantone colors

The inside of this box is printed on paper and has a design pattern on all sides. It uses just one color—the patterns are white, the natural color of the paper. CorelDraw was used to create this design.

DESIGN FIRM revoLUZion
ALL DESIGN Bernd Luz
CLIENT Farmer Spiess
PRODUCT Farm food
TECHNIQUE Watercolor, ink, offset

The pictures are watercolor, and icons for the products were made with a brush. The images were manipulated in Adobe Photoshop and QuarkXPress and printed on a natural paper.

DESIGN FIRM FLB Design Limited
ART DIRECTOR Colin Mechan, Robin Bennion
DESIGNER Colin Mechan
PHOTOGRAPHER Eric Mandel
CLIENT Asda Stores Ltd.
PRODUCT Hot Eating Pies
TECHNIQUE Cartons–lithography, film packs– flexography

These designs focus on the orange vignetted bands on the left side of each pack to imbue warmth and flavor. Simplistic badging details the pastry type and product variety.

food | 87

DESIGN FIRM
FLB Design Limited
ART DIRECTOR/DESIGNER
Colin Mechan
CLIENT
The Jacob's Bakery
PRODUCT
Twiglets
TECHNIQUE
Bags–flexography,
caddy labels–lithography

The client wanted its Twiglets packaging to reflect the "uniquely knobbly" snack, taking the brand into the "Twilight Zone." The packaging contains product photography of the twiglets combined with digitally generated illustrations of the vortex element. Final artwork was produced in Adobe Illustrator.

DESIGN FIRM
FLB Design Limited
ART DIRECTOR/DESIGNER
Susan Fosbery
PHOTOGRAPHER
Andy Seymour
CLIENT
Terry's Suchard
PRODUCT
Continental Selection chocolates
TECHNIQUE
Lithography

A marbled, full-color background illustration was commissioned, and individual chocolates were photographed with bright, interesting colored papers. Gold foil blocking was used to enhance the premium cues. Master hand-lettering was conventionally created for Continental Selection. Full digital artwork was produced by using Adobe Illustrator.

88 | Package & Label Design

DESIGN FIRM
The Design Company
ART DIRECTOR
Marcia Romanuck
DESIGNER
Fran McKay
CLIENT
Metropolis Fine Confections
PRODUCT
Metropolis Bark products
TECHNIQUE
Offset

The chocolates were digitally photographed, scanned in Adobe Photoshop, and placed in QuarkXPress. The textured backgrounds were created in Adobe Illustrator and combined with the images in Photoshop. These were printed in four-color process on a recycled woven stock.

DESIGN FIRM
The Design Company
ART DIRECTOR/DESIGNER
Marcia Romanuck
CLIENT
Dillon Candy Company
PRODUCT
Dillon pecan and peanut candy
TECHNIQUE
Flexography

This project involved redesigning the logo, packaging, and labels, while updating the company's previous mom-and-pop image yet still emphasizing its southern heritage. These packages were printed in four PMS colors, including a metallic gold. The art was created in Macromedia FreeHand.

food | 89

DESIGN FIRM Lewis Moberly
ART DIRECTOR Mary Lewis
DESIGNER Paul Cilia La Corte
ILLUSTRATOR Geoff Appleton
CLIENT Askeys
PRODUCT Cup Cornets

The design's focus is on the pleasure aspect of ice cream while evoking happy childhood memories. The packaging has dual merchandising, which means it can be displayed as a portrait or landscape. The bright bold posterised imagery links up on a shelf to create a continuous Askey's world.

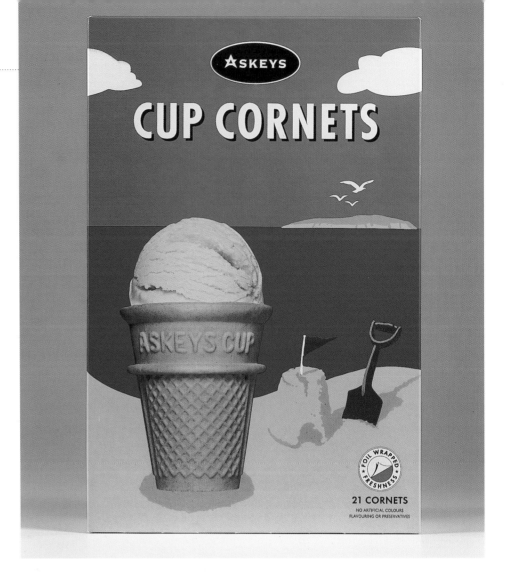

DESIGN FIRM Sayles Graphic Design
ALL DESIGN John Sayles
CLIENT Hickory Pit
PRODUCT Hickory Pit Barbecue Sauce
TECHNIQUE Offset

These labels feature the designer's original graphics with warm colors that stand out in retail displays. The restaurant's distinctive logo, also designed by Sayles, is wrapped in flames and smoke, reminiscent of a rotisserie grill. Additional graphics on the label are illustrated in the same style.

DESIGN FIRM Bartels & Company, Inc.
ART DIRECTOR David Bartels
DESIGNER Aaron Segall
ILLUSTRATOR Mary Grand Pré
CLIENT Hank's Cheesecake
PRODUCT Cheesecake
TECHNIQUE Four-color and varnish

Building on a theme, Bartels & Company, Inc. used an angel motif and one black feather that says eating a Hank's cheesecake is only a small sin. In total, three different size boxes were produced and have been met with enthusiastic acceptance.

DESIGN FIRM
Richards Group Inc.
ART DIRECTOR
Debbie Crawford
DESIGNER
Laura Kuhlman
ILLUSTRATOR
Alan Kastner
CLIENT
Great Dakotas Baking Co.
PRODUCT
Baked goods

The firm designed the packaging, promotional materials, and point-of-purchase display. An excellent consumer response extended the product line to include twenty-five bread items, cinnamon rolls, gourmet cookies, and sandwiches.

food | 91

DESIGN FIRM Hornall Anderson Design Works
ART DIRECTOR Jack Anderson
DESIGNER Jack Anderson, Jana Nishi, Mary Hermes, Heidi Favour, David Bates, Mary Chin Hutchison
ILLUSTRATOR Keith Ward
CLIENT Seattle Chocolate Company
PRODUCT Seattle Chocolates

A family look was created by developing a black corrugated bottom to be used on all flavors. A combination of tip-ins, hot-stamping debossing, embossing, and matte and gloss varnishes were used to give the entire package the special feeling it needed for a gift box.

DESIGN FIRM Greteman Group
ART DIRECTOR Sonia Greteman, James Strange
DESIGNER/ILLUSTRATOR James Strange
CLIENT Menefee & Partners
PRODUCT Pizza

This fun Cafe Doskocil pizza box was used as a promotion to sell pizza toppings. The art is retro, using black and red for impact. It was created in Macromedia FreeHand.

DESIGN FIRM Hornall Anderson Design Works
ART DIRECTOR Jack Anderson
DESIGNER Jack Anderson, David Bates
ILLUSTRATOR David Bates
CLIENT Alki Bakery
PRODUCT Alki Bakery packaging

To achieve the desired effect, a logo was developed to depict the bakery's old-world home-style of baking through the illustration of a wooden bread peel being placed in an oven.

DESIGN FIRM Hornall Anderson Design Works
ART DIRECTOR Jack Anderson
DESIGNER Jack Anderson, John Anicker
ILLUSTRATOR John Fretz
CLIENT Continental Mills
PRODUCT Krusteaz Bread Machine Mix package

The packaging consists of a warm color palette with illustrations emphasizing a natural, old-world, home-style appeal. Within a line, individual product flavors are clearly differentiated through a family of colors with one predominant color applied to each.

DESIGN FIRM Del Rio Diseño
ALL DESIGN Del Rio Diseño Team
CLIENT Costa
PRODUCT Costa-Cookies—Costa Fiesta
TECHNIQUE Rotogravure

The entire family of Costa cookies has a band that highlights the brand Fiesta to reinforce the unity of the line. Each variety of cookie is distinguished by a different color.

DESIGN FIRM Sibley-Peteet Design
ART DIRECTOR Don Sibley
DESIGNER/ILLUSTRATOR John Evans
CLIENT Frito Lay
PRODUCT Ruffles potato chips
TECHNIQUE Computer

Frito Lay put Ruffles potato chips in an aluminum can with a pull top so consumers could throw it in a backpack and the chips would not get crushed. This design was done in Adobe Illustrator.

DESIGN FIRM Lambert Design
ART DIRECTOR Christie Lambert
DESIGNER/ILLUSTRATOR Joy Cathey Price
CLIENT Duo Delights
PRODUCT Reindeer Feed (candy)
TECHNIQUE Offset

Reindeer Feed is white chocolate and nuts and is marketed during the holidays in catalogs and specialty stores. The concept was to make a hang tag that resembles a deer crossing warning sign, but with a red-nosed reindeer. All the art was created in Adobe Illustrator.

DESIGN FIRM Animus Comunicação
ART DIRECTOR/DESIGNER Rique Nitzsche
ILLUSTRATOR Arturo Uranga
CLIENT Nabisco
PRODUCT Christmas box biscuits
TECHNIQUE Four-color process

Nabisco usually creates different temporary packages for special holidays to help the product stand out at point-of-purchase. This Christmas box reunites four Nabisco products. The consumer will recognize the products printed on the top of the box.

DESIGN FIRM Animus Comunicação
ART DIRECTOR Rique Nitzsche
DESIGNER Rique Nitzsche, Felicio Torres
PHOTOGRAPHER Studio Oficina
CLIENT Nabisco
PRODUCT Cream cracker
TECHNIQUE Laminating, offset

This package needed to be easily printed because consumer volume is very high. Some months later, the firm designed the count-unit package for the same product and a shipping package that served as an exhibition display.

DESIGN FIRM Animus Comunicação
ART DIRECTOR Rique Nitzsche
DESIGNER Rique Nitzsche,
 Claudia Lobão
ILLUSTRATOR Eliane Soares
CLIENT Fleischmann Royal
PRODUCT Pommy's
TECHNIQUE Offset

Pommy's is a complete line of homemade-style jams, preserves, and sauces. The idea was to create an old-fashioned look with hand-lettering titles over a white label to contrast with the product's normally dark color inside the goblet.

DESIGN FIRM Animus Comunicação
ART DIRECTOR Rique Nitzsche
DESIGNER Rique Nitzsche,
 Sandra Ribeiro
PHOTOGRAPHER Al Handan
CLIENT Nabisco
PRODUCT Chipits
TECHNIQUE Offset

Nabisco asked Animus to develop a new package targeting young consumers. As a result, Animus used strong colors to contrast with the previous "colorless" packaging. The new packaging has helped increase the product's sales.

food

DESIGN FIRM Goodhue & Associés
ART DIRECTOR Suzanne Côté
DESIGNER Suzanne Côté, Nicole Bouchard
PHOTOGRAPHER Louis Prud'homme
CLIENT Épiciers unis Métro-Richelieu
PRODUCT Granola cereal

This packaging reflects the nutritional value of the product by contrasting backgrounds to set low-fat cereals apart from regular brands. The colors of the packaging are striking.

DESIGN FIRM Goodhue & Associés
ART DIRECTOR Suzanne Côté
DESIGNER Suzanne Côté, Nicolas Jouhannaud
ILLUSTRATOR Marie Lessard
CLIENT Épiciers unis Métro-Richelieu
PRODUCT Pasta

A plethora of garden vegetables surrounds a window that reveals the type of pasta enclosed. Vivid colors on a white background mean the product is sure to stand out.

DESIGN FIRM Goodhue & Associés
ART DIRECTOR/DESIGNER Nathalie Houde
ILLUSTRATOR Louis Prud'homme
CLIENT Épiciers unis Métro-Richelieu
PRODUCT Grano Fruit (fruit bar)

The concept of this packaging revolves around a relaxed family ambiance and consumer concern for wholesome, balanced nutrition. The product's name reflects these values.

DESIGN FIRM
Goodhue & Associés
ART DIRECTOR
Suzanne Côté
DESIGNER
Suzanne Côté, Nicolas Jouhannaud
ILLUSTRATOR
Atelier du Presse-Citron
PHOTOGRAPHER
Louis Prud'homme
CLIENT
Épiciers unis Métro-Richelieu
PRODUCT
Croutons

A rustic tabletop covered with vegetables fresh from the garden and ready to become a delicious country-style meal entice the consumer.

food | 99

DESIGN FIRM Pinkhaus
ART DIRECTOR Todd Houser, Joel Fuller
DESIGNER Todd Houser
ILLUSTRATOR David Diaz
CLIENT Fiesta Pop
PRODUCT Popcorn
TECHNIQUE Screen

The client wanted packaging that would stand out from the competition, so Pinkhaus used an attractive "illustrative" look and gave the background a Latin fiesta feeling. It was done in Adobe Illustrator and printed at a third world screen printer.

DESIGN FIRM
CMA
ART DIRECTOR
Bob Milz
DESIGNER/ILLUSTRATOR
Trish Hill
CLIENT
Lou Ana
PRODUCT
Lou Ana Fish Fry
TECHNIQUE
Flexography

The design for this breading/coating mix communicates the product through a simple and colorful fish illustration. The illustration and finished production art for the package were created in Macromedia FreeHand.

100 | Package & Label Design

DESIGN FIRM Mike Salisbury Communications
ART DIRECTOR/DESIGNER Mike Salisbury
ILLUSTRATORS Bruce Eagle, Pat Linse
CLIENT RJR Nabisco
PRODUCT Bubble Yum Bubble Gum
TECHNIQUE Airbrush

Bubble Yum came to Mike Salisbury to try to recapture market share that they were losing to the competition. The old packaging was invisible on the racks, so the redesign used neon colors.

Beverages

DESIGN FIRM Primo Angeli Inc.
ART DIRECTOR Primo Angeli,
    Ron Hoffman
DESIGNER Jenny Baker, Mark Jones
ILLUSTRATOR Liz Wheaton
CLIENT Gourm-E-Co. Imports
PRODUCT Herbal tea

The goal for this assignment was to create an herbal line of imported English teas with a perceived established heritage. The labels were proposed to be run on a mix-and-match single set of colors on one press run. Labels were then hand-applied to a basic black stock item bag with a see-through window.

DESIGN FIRM Primo Angeli Inc.
ART DIRECTOR Primo Angeli
DESIGNER Mark Jones
STRUCTURAL DESIGN Kornick Lindsay (Chicago)
CLIENT Coca-Cola USA
PRODUCT Coca-Cola

The designer worked with the Chicago structural design firm Kornick Lindsay to explore a range of graphic and structural possibilities for a two-piece and three-piece can. Coca-Cola management ultimately decided on a structure that emulated the shape of its famous glass bottle because of consumer identification. After much graphic exploration, the designer's simplest adaptation of the existing Coca-Cola equities was used.

DESIGN FIRM
Primo Angeli Inc.
ART DIRECTOR
Carlo Pagoda, Primo Angeli, Brody Hartman
DESIGNER
Mark Jones, Brody Hartman
ILLUSTRATOR
Rick Gonella
TYPOGRAPHER
Sherry Bringham
CLIENT
G. Heileman Brewing Co.
PRODUCT
Henry Weinhard's root beer

Primo Angeli Inc. proposed a long-neck brown beer bottle for the package, but the company thought that type of bottle should be used only for beer. Within the context of the total package, the long-neck bottle changed its beer association.

DESIGN FIRM
Primo Angeli Inc.
ART DIRECTOR
Primo Angeli, Rolando Roster
DESIGNER
Terrence Tong, Brody Hartman, Paul Terrill
ILLUSTRATOR
Peter Sin
CLIENT
Nestlé Beverage Company
PRODUCT
Sarks Coffee

A warm background of rich bronzes and creams evokes the richness of the coffee while the Sarks name, in bold, elegant type, is placed high up on the package for increased recognition and visibility. For heightened texture, the Café Sarks statement of quality runs in subtly toned italic type around the package.

beverages | 105

DESIGN FIRM
Primo Angeli Inc.
ART DIRECTOR
Primo Angeli
DESIGNER
Primo Angeli,
Doug Hardenburgh
CLIENT
Pete's Brewing Company
PRODUCT
Beer

Upgraded packaging for Pete's Wicked Ale projects a premium microbrewery identity with the help of foggy shipyard images that evoke the fine ales enjoyed by serious ale-drinking regulars at a favorite pub. The project included a packaging graphics redesign and the addition of two new flavors.

DESIGN FIRM
Primo Angeli Inc.
ART DIRECTOR
Jerry Andelin (Hal Riney + Partners), Primo Angeli
DESIGNER/TYPOGRAPHER
Mark Jones
ILLUSTRATOR
Bruce Wolfe
CLIENT
Hal Riney + Partners
PRODUCT
Beer

The objective was to create an identity for a regional microbrewery.

106 | Package & Label Design

DESIGN FIRM Tieken Design and Creative Services
ART DIRECTOR Fred E. Tieken
DESIGNER Fred E. Tieken, Rik Boberg
CLIENT Black Mountain Brewing Company
PRODUCT Juanderful Weizen beer
TECHNIQUE Offset

Because wheat beer was originally brewed by monks, the client wanted to convey an old-world, somewhat monastic message through the graphics. The woodcut-style illustrations of shocks of wheat and barrel add to the "made-by-hand" appeal. The lettering and graphics were created in Adobe Illustrator and imported into Adobe Photoshop and manipulated.

DESIGN FIRM Carter Wong and Partners Ltd.
ART DIRECTOR Philip Carter
DESIGNER Philip Carter, Teri Howes
CLIENT AC Water Canada Inc.
PRODUCT Crystal Canadian
TECHNIQUE Four-color process, lithography

The Maple Leaf was used to convey the Canadian origins with the expressed permission of the Canadian government. The simplicity of the color-coded label design illustrates the crisp, pure qualities of the water. Silver edging enhances the premium nature of the product. The label was created in Adobe Illustrator.

beverages | 107

DESIGN FIRM FRCH Design Worldwide
ART DIRECTOR Joan Donnelly
DESIGNER Tim A. Frame
CLIENT Borders Books & Music
PRODUCT Coffee
TECHNIQUE Flexography

The artwork for this label was created in Adobe Illustrator. Labels were color coded for regular, decaffeinated, and flavored varieties with designated areas for write-in information such as weight and date sold.

DESIGN FIRM Tieken Design and Creative Services
ART DIRECTOR Fred E. Tieken
DESIGNER Rik Boberg, Fred E. Tieken
CLIENT Black Mountain Brewing Company
PRODUCT Chili Light Beer
TECHNIQUE Label–offset, six-pack carrier–flexography

The client wanted to convey the same vibrant and colorful Southwestern image of its orginal product, so the designers used clouds and floating chilies to convey the "lightness" of this new product. The lettering and graphics were created in Adobe Illustrator and then merged with a CD-ROM photo image of clouds in Adobe Photoshop.

DESIGN FIRM Tharp Did It
ART DIRECTOR Rick Tharp, David Hansmith
DESIGNER/ILLUSTRATOR Rick Tharp
CLIENT Mount Konocti Winery
PRODUCT Mount Konocti Pear Brandy
TECHNIQUE Letterpress

The artwork for this label was handmade, as is the brandy. The label was letterpressed in four colors and applied by hand.

beverages | 109

DESIGN FIRM Sommese Design
ART DIRECTORS Kristin Sommese, Lanny Sommese
DESIGNER Lanny Sommese
CLIENT AquaPenn Spring Water Co.
PRODUCT Black Panther Spring Water
TECHNIQUE Offset

The purpose of the design is perhaps best described on the back of the label, "black panther's eyes have captured you." The panther connotes perfection, style, purity, and excellence. The eyes of the cat were designed to confront buyers from shelf.

DESIGN FIRM fotodesign + conception
ALL DESIGN Werner Quester
CLIENT fotodesign + conception
PRODUCT wine

These labels are used for different wines that fotodesign + conception gave to clients as holiday gifts. The designer creates a new one every year.

DESIGN FIRM Watts Graphic Design
ART DIRECTOR/DESIGNER Helen and Peter Watts
CLIENT Watts Graphic Design
PRODUCT Christmas Gift
TECHNIQUE Laser print with foil

Every year Watts Graphic Design gives its clients a gift to show appreciation. This self-promotion was a great success. Clients called to comment on the unique design. It is simple but effective.

beverages | 111

DESIGN FIRM Barbara Ferguson Designs
ALL DESIGN Barbara Ferguson
CLIENT San Diego Zoo
PRODUCT Kids' polar bear bucket and cup
TECHNIQUE Screen

The San Diego Zoo is proud of its new, more natural environment for its polar bear exhibit. The bears seem happy, too. This children's bucket and cup design depicts the playfulness and activity found by all who visited the bears' new home.

DESIGN FIRM Barbara Ferguson Designs
ALL DESIGN Barbara Ferguson
CLIENT San Diego Wild Animal Park
PRODUCT Beverage cup
TECHNIQUE Screen

The client was looking for not only a fun design for both adults and children, but also a design that depicted the unusual alligator in its environment. This reversed-out format has been an extreme success and was later applied to stickers, shirts, patches, and magnets.

DESIGN FIRM Del Rio Diseño
ART DIRECTOR/DESIGNER Del Rio Diseño team
ILLUSTRATOR Rodrigo Vega
CLIENT Viña Concha y Toro
PRODUCT Bag-in-box Santa Emiliana
TECHNIQUE Offset

This bag-in-box for wines recalls the origin of the product through an impressionistic pictorial treatment that produces a strong visual impact when the boxes are stacked.

DESIGN FIRM Del Rio Diseño
ART DIRECTOR/DESIGNER Del Rio Diseño team
CLIENT Viña Carta Vieja
PRODUCT Bag-in-box Carta Vieja
TECHNIQUE Offset

A label as the main design element for this bag-in-box conveys a traditional concept of wine. The wine glass and tavern from the Renaissance period were integrated in order to suggest a consuming environment.

beverages | 113

DESIGN FIRM Marsh, Inc.
ART DIRECTOR Greg Conyers, Ken Neiheisel
DESIGNER Greg Conyers
PHOTOGRAPHER Bill Magness, Ed Betz
CLIENT Industrias Banilejas
PRODUCT Induban ground coffee
TECHNIQUE Offset

The objective here was to design an upscale anniversary package for an existing Santo Domingo coffee. The package was created in QuarkXPress, incorporating the brand's current logotype.

DESIGN FIRM Daniel K. Brown
ART DIRECTOR/DESIGNER Daniel K. Brown
PHOTOGRAPHER Richard Scanlan
CLIENT Saroc, Inc.
PRODUCT Valley of the Moon natural spring water from Montana
TECHNIQUE Offset

This product label is a full wrap-around series of black mountains with gold lettering, which was designed in QuarkXPress. The blue bottle is cast with a raised crescent moon centered between the mountain peaks, making the bottle itself part of the label concept.

114 | **Package & Label Design**

DESIGN FIRM
Pentagram Design Inc.
ART DIRECTOR
Woody Pirtle
DESIGNER
Woody Pirtle, Patricia Choi
ILLUSTRATOR
Woody Pirtle
CLIENT
Flying Fish Brewing Co.
PRODUCT
Ale
TECHNIQUE
Offset

Instead of backward-looking imagery—sheaves of barley, revolutionary war heroes—the client wanted something unabashedly modern. The whimsical Flying Fish label came out in 1995; it signifies a different, edgier product.

beverages | 115

DESIGN FIRM Studio Associates
ART DIRECTOR/DESIGNER Phil Gajewski
ILLUSTRATOR Durwood Coffey
CLIENT Stoney Creek Brewing Company
PRODUCT Stoney Creek Michigan Lager
TECHNIQUE Mixed media

This packaging was designed to convey a midwestern, Michigan feel. It was produced using a variety of media, such as photography, conventional illustration, and digital design/layout.

DESIGN FIRM The Coleman Group
ART DIRECTOR Karen Corell
DESIGNER Carlos Seminario
ILLUSTRATOR Michael Wepplo
CLIENT Tropicana Products, Inc.
PRODUCT Tropicana Pure Premium orange juice
TECHNIQUE Offset

As a result of consumer research, the original design architecture was maintained with enhancements to the Tropicana logo, the fruit illustration, and the product designator on the new packaging. The ribbon element was used as a branding device for Pure Premium, and the "Not From Concentrate" communication was given more prominence.

116 | Package & Label Design

DESIGN FIRM Hans Flink Design Inc.
ART DIRECTOR Hans D. Flink
DESIGNERS Denise Heisey, Hans D. Flink
ILLUSTRATORS Julia Noonan, Jaque Auger
CLIENT Cadbury Beverages, Inc.
PRODUCT Mott's juice line

The labels for this large, new line of juice drinks needed to feature a strong banding system, a modern setting, and bold, realistic images for flavor appeal. Airbrush illustrations and hand-rendered script was combined with Macintosh-produced art files.

DESIGN FIRM The Coleman Group
ART DIRECTOR Edward Morrill
DESIGNER John Rutig, Carlos Seminario
CLIENT Brown-Forman Beverage Co.
PRODUCT Early Times
TECHNIQUE Offset, embossing, gold stamping

While maintaining equity with the existing packaging image, these new graphics enhance the use of warm rich colors, refined Early Times logotype, shoulder seal stamp of quality, and higher quality typography throughout.

beverages | 117

DESIGN FIRM Design Resources
ART DIRECTOR Ken Schwager
DESIGNER Ken Schwager, Dick Tuttle
ILLUSTRATOR Craig Adams
CLIENT Beverage House
PRODUCT Ready-Bru tea
TECHNIQUE Flexography, six colors

The design objectives were to create label designs that were flexible enough to be used on more than one flavor and conformed to sizes and formats required by the client's existing labeling equipment. The design was created using QuarkXPress, Macromedia FreeHand, and Adobe Photoshop. The tea splashing was a composite of ten photos with digital retouching. Plastic caps were colored to match the label colors.

DESIGN FIRM Design Source, Inc.
ART DIRECTOR/DESIGNER Robert Salazar
PHOTOGRAPHER Edgar Bayani
CLIENT Coco Manila Food Corp.
PRODUCT Juko (drink)
TECHNIQUE Offset

Juko is the first package Tetra Pak sent to Sweden for printing in diskette form. It was composed in Macromedia FreeHand, and it consists of five language translations on the back. The packaging received praise from International Food Fairs.

DESIGN FIRM Washam Design
ART DIRECTOR Thurlow Washam
DESIGNER Thurlow Washam, Annemarie Clark
ILLUSTRATOR Thurlow Washam
CLIENT J. Fritz Winery
PRODUCT Wine
TECHNIQUE Offset, foil stamp

The client wanted to update its heavy European look to a lighter Californian feel. The logo represents the client's three doors at the entrance of the cellar, and customized its colors to coordinate with the fruit flavor of each variety. These colors are used on all collateral material pertaining to each wine. It was created in QuarkXPress and Adobe Illustrator.

beverages | 119

DESIGN FIRM DIL Consultants in Design
ALL DESIGN DIL Staff
CLIENT Cia. Cerveharia Brahma
PRODUCT Brahma soft drinks
TECHNIQUE Dry-offset printing

Aiming to promote its line of soft drinks during the months before the Olympic Games, Brahma solicited the designers to develop promotional cans. Each brand (Gurana, Sukita, and Limao) received a different design of an Olympic sport, especially the ones which had best chances of winning a medal for Brazil.

DESIGN FIRM Clark Design
ART DIRECTOR Annemarie Clark
DESIGNER/ILLUSTRATOR Thurlow Washam
CLIENT North Coast Coffee
PRODUCT Coffee
TECHNIQUE Offset

The client requested an upscale look with more sophistication than a coffee boutique, so Clark Design used metallic silver to add elegance and to match the silver bags that had to be used in packaging. It was created in QuarkXPress and Adobe Illustrator.

120 | Package & Label Design

DESIGN FIRM
CMA
ART DIRECTOR
Bob Milz
DESIGNER
Leon Alvarado
CLIENT
Coors Brewing Co.
PRODUCT
George Killian's Wilde Honey
TECHNIQUE
Macromedia FreeHand

The label and six-pack design relies on a honeycomb pattern. Warm, honey-gold coloration was added to the traditional architecture of the Killian's label to communicate the sweet difference in the flavor of the ale inside.

DESIGN FIRM
CMA
ALL DESIGN
Bob Milz
CLIENT
Texas Coffee Co.
PRODUCT
Seaport Coffee
TECHNIQUE
Macromedia FreeHand

This updated design is printed directly on foil, with a new, deeper palette of colors for a more dramatic look on grocery shelves. Elements of the original Seaport coffee label design include the coffee freighter, sea gull, and the large serpentine S of the Seaport logotype.

beverages | 121

DESIGN FIRM CMA
ART DIRECTOR/DESIGNER Leon Alvarado
PHOTOGRAPHER Ralph Smith Photography/Greg Dawson
CLIENT Coca-Cola Foods
PRODUCT Minute Maid Soft Frozen Lemonade
TECHNIQUE Flexography

Minute Maid Soft Frozen Lemonade was created, packaged, and produced within 230 days prior to the 1996 Olympic Games specifically in order to take advantage of the Atlantic Olympic market. CMA extended the line look of Minute Maid products, showing a photographic spoonful of the product for clear identification and taste appeal.

DESIGN FIRM CMA
ART DIRECTOR/DESIGNER Leon Alvarado
PHOTOGRAPHER Ralph Smith Photography/Greg Dawson
CLIENT Coors Brewing Co.
PRODUCT Coors Cutter
TECHNIQUE Macromedia FreeHand

Focus group research showed that consumers of no- and low-alcohol products wanted to look like they were drinking regular beer in a crowd of other drinkers. CMA adapted the traditional heraldry of Coors Brewing, reviving an antique version of the Coors Castle Rock logo.

DESIGN FIRM Heather Sumners
ALL DESIGN Heather Sumners
CLIENT Bender's Beverage Co.
PRODUCT Non-alcoholic grape juice
TECHNIQUE Color pencil, typesetting

This label design for Bender's non-alcoholic grape juice was produced and created in QuarkXPress. The client wanted a clean, simple, classic design that would be timeless. The use of muted classic colors helped achieve that effect.

DESIGN FIRM JCHO
ALL DESIGN Juan Carlos, Hernández Ortega
CLIENT Mezcal JCHO
PRODUCT Mezcal
TECHNIQUE Paper, ink, rope, bottle

The designers glued rope around a bottle with a cork. They designed the label, cut it out, and glued it to the bottle, then designed a logo for the cork.

beverages | 123

DESIGN FIRM Steep
ART DIRECTOR/DESIGNER Jeff Piazza
DESIGNERS/MECHANICALS Josh Berger, Jean Innocent, Niko Courtelis, Damion Silver
PHOTOGRAPHER Kevin Thomas
CLIENT Steep
PRODUCT Steep Tea Matchbox Display Unit
TECHNIQUE Offset

The trademarked matchbox design holds the large whole-leaf tea sack and has a secret compartment for a hidden message. To save on cost, the designer used standard boxes with almost twenty different labels. The chipboard display units hold fifteen tea or cocoa matchboxes, are printed in black and silver, and have text describing the company's philosophy hidden throughout the inside of the box.

DESIGN FIRM Robilant & Associati
ART DIRECTOR Maurizio di Robilant
DESIGNER Lucia Sommaruga
ILLUSTRATOR Nicola Mincione
CLIENT illycaffé Spa Italy
PRODUCT MITE coffee

The word Mite means mild. This coffee is obtained by selecting naturally light coffees. The aim of this packaging is to clearly communicate an idea of nature and mildness. Mild and mat colors and the selection of mat paper support this effect.

124 | Package & Label Design

DESIGN FIRM
Robert Bailey Incorporated
CREATIVE DIRECTOR
Robert Bailey
ART DIRECTOR/DESIGNER
Dan Franklin
ILLUSTRATOR
Katie Doka
CLIENT
Boyd Coffee Company
PRODUCT
Flavored syrups
TECHNIQUE
Flexography

Flavored syrups are part of traditional Italian feasts and celebrations. Italia D'Oro is a logo and brand identity redesign to support the product upscale acceptance. It was designed in Macromedia FreeHand and illustrated by hand.

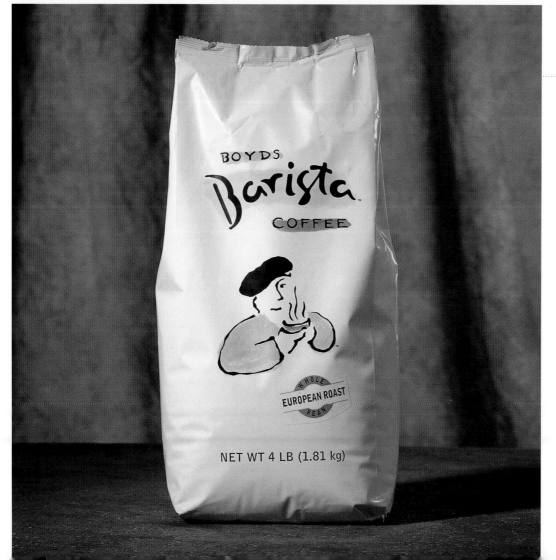

DESIGN FIRM
Robert Bailey Incorporated
CREATIVE DIRECTOR
Robert Bailey
ALL DESIGN
Dan Franklin
CLIENT
Boyd Coffee Company
PRODUCT
Coffee
TECHNIQUE
Flexography

Boyd's Barista coffee is like the chef's cut of coffee. This design was great for the flamboyant Barista, introduced in November 1995. The product is setting sales records. The packaging was designed and illustrated in Macromedia FreeHand.

beverages | 125

DESIGN FIRM Mike Salisbury
   Communications
ART DIRECTOR/DESIGNER Mike Salisbury
CLIENT Suntory
PRODUCT Risque liqueur
TECHNIQUE Fabric, metal

Suntory wanted to position this product as a seductive after-dinner drink targeted at a younger, but sophisticated drinker. Mike Salisbury Communications made a jacket of black vinyl with a big zipper which was later mass produced.

DESIGN FIRM DesignCentre of Cincinnati
ART DIRECTOR Mark Castator
DESIGNER Joan Bishop Gottsacker
CLIENT The Kroger Co.
PRODUCT Sports Shot
TECHNIQUE Flexography

Sports Shot is targeted at young men. In the redesign, DesignCentre focused on fun, energy, and flavor rather than the activity specific, health aspects of the beverage in order to make an impact.

126 | Package & Label Design

DESIGN FIRM Mires Design
ART DIRECTOR John Ball
DESIGNER John Ball, Miguel Perez
ILLUSTRATOR Tracy Sabin
CLIENT Heartford Dairy/Ken C. Smith
PRODUCT Lowfat milk
TECHNIQUE Flexography

This product is a heart-healthy lowfat milk. The packaging was designed for flexographic printing with bold, solid areas of color and limited registration.

DESIGN FIRM Mires Design
ART DIRECTOR Mike Brower
DESIGNER Mike Brower, Deborah Hom
ILLUSTRATOR Tracy Sabin
CLIENT Food Group
PRODUCT Blended coffee mix

Mires Design felt that in order to give this new product more credibility, it should create the packaging with a consumer approach. Thus, this packaging is more appealing to the eye as opposed to other commercial packaging.

beverages | 127

DESIGN FIRM
FLB Design Ltd.
ART DIRECTOR
Colin Mechan
DESIGNER
Susan Fosbery
ILLUSTRATOR
Ian Mudie
CLIENT
Kraft Jacobs Suchard
PRODUCT
Maxwell House Instant Coffee Powder
TECHNIQUE
Lithography

An initial black-and-white illustration was produced to emulate the background swirl effect—this was then scanned into a Scitex system and manipulated to work around the type. A full-color illustration of the coffee beans was created to enhance the taste cues. Full digital artwork was produced using Adobe Illustrator. FLB then project managed the origination stage, which included production of films and wet proofs.

DESIGN FIRM
Westpoint Stevens Inc. design group
ART DIRECTOR
Gail Rigelhaupt
DESIGNER
Risa Brand
CLIENT
Westpoint Stevens Inc.
PRODUCT
Tea
TECHNIQUE
Offset

The labels of this promotional gift were developed using patterns from four of the company's major bedding collections. Each tea was chosen to relate to the concept of the collection (i.e., English Breakfast for a collection adapting a William Morris print, Chamomile, a calming tea, for the "weekend," and so on).

128 | Package & Label Design

DESIGN FIRM
Karacters Design Group
ART DIRECTOR
Matti Chan
CREATIVE DIRECTOR
Maria Kennedy
CLIENT
Nabob Coffee Company
PRODUCT
Nabob canisters
TECHNIQUE
Screen

A promotional program involving the purchase and redemption of UPC codes for high-quality designer canisters reinforced Nabob's dedication to its customers. Dramatic line-art illustration used on the coffee packaging itself and tasteful ornate detailing, gave an overall line look to the canisters, which proved to be a true collector's edition.

DESIGN FIRM
FLB Design Limited
ART DIRECTOR/DESIGNER
Colin Mechan
PHOTOGRAPHER
Eric Mandel
CLIENT
Kiri
PRODUCT
Appletise
TECHNIQUE
Lithography

Transparencies of the apples, currants, oranges, and face image were scanned into a Mamba system, and color correction and manipulation helped to comp together the face image, fruit, and color wash and enable final output of all graphics for each variety as one transparency. Final digital artwork was supplied with transparencies to the printers.

beverages | 129

DESIGN FIRM Karacters Design Group
ART DIRECTOR Jeff McLean
CREATIVE DIRECTOR Maria Kennedy
ILLUSTRATOR Jeff McLean
CLIENT Clearly Canadian Beverage Corp.
PRODUCT Orbitz
TECHNIQUE Screen

The elegant proprietary bottle shape recalls a lava lamp, highlighting the product inside. The use of APL label and coordinating bright colors of the logo allowed the product to be seen through the label. The final product has captured the attention of the teen market, urging them to "Defy Gravity" and "Get into Orbitz."

DESIGN FIRM Bunny Levy & Associates
ART DIRECTOR Bunny Levy, Paulette Lue
DESIGNER Paulette Lue
ILLUSTRATOR Cathy Diefendorf
  (Mendola Artists rep)
CLIENT Desnoes & Geddes Ltd.
PRODUCT Hampden Club Rums
TECHNIQUE Offset

Jamaica has a long tradition of rum making. As a new player, Desnoes and Geddes Ltd. sought to attract a new and younger rum drinking audience with a dynamic, contemporary design. Aldus Freehand and Adobe Photoshop were used.

DESIGN FIRM Bunny Levy & Associates
ART DIRECTOR Bunny Levy, Paulette Lue
DESIGNER Paulette Lue
ILLUSTRATOR Cathy Diefendorf,
  Joyce Kitchell, Larry Winborg
CLIENT Estate Industries
PRODUCT Lillifield liqueur
TECHNIQUE Offset

Illustrators with loose, painterly styles were chosen for this project. Each illustrator was given freedom with composition, color, and illustrative style. Embossing was used to further enhance areas of the illustrations as well as the typeface. Aldus Freehand and Adobe Photoshop were used.

beverages | 131

DESIGN FIRM The Design Company
ART DIRECTOR Marcia Romanuck
DESIGNER/ILLUSTRATOR Marcia Romanuck, Busha Husak
CLIENT Habersham Winery
PRODUCT Habersham wine
TECHNIQUE Offset, four-color PMS

This was an existing product line that needed a more sophisticated look to match the quality of the wine. The art was illustrated by hand and scanned into a Quark document.

DESIGN FIRM Lewis Moberly
ART DIRECTOR/DESIGNER Mary Lewis
CLIENT Arnold Dettling
PRODUCT Dettling Kirschwasser

Kirsch typically is packaged in clear flint standard bottles, but Dettling wanted to break the mold and design a unique bottle. The colored glass reflects the deep stain of black cherries. The slim bottle has a flat back and curved front with an applied label only on the neck. The brand's credentials are in hand-drawn clear script against a frosted surface.

DESIGN FIRM Sayles Graphic Design
ALL DESIGN John Sayles
CLIENT Timbuktuu Coffee Bar
PRODUCT Coffee Beans-To-Go packaging
TECHNIQUE Offset

Carry-out packaging and in-store retail packaging carry Sayles' signature "bean man" and tribal mask graphics developed for the Timbuktuu Coffee Bar.

DESIGN FIRM Saatchi & Saatchi Denmark
ART DIRECTOR/DESIGNER Vibeke Nodskov
CLIENT Wiibroe
PRODUCT Natur Pils (Eco beer)
TECHNIQUE Offset

The label for this new Eco beer was made in QuarkXPress. An old original engraving was used as illustration and printed in mat gold. Brown wrapping paper was scanned to give structure and texture to the background for a natural, simple look and quality.

DESIGN FIRM Greteman Group
ART DIRECTOR Sonia Greteman
DESIGNER Sonia Greteman, Chris Parks, Craig Tomson
CLIENT Oaxaca Grill
PRODUCT Restaurant menu, label, and wine list

The Oaxaca Grill menu, wine bottle label, and wine list add the flavor of Mexico to this authentic restaurant. The Greteman Group used symbolic imagery of the region of Oaxaca and combined it with a rustic metal binding and a washable paper choice. It was created in Macromedia FreeHand.

DESIGN FIRM Hornall Anderson Design Works
ART DIRECTOR Jack Anderson
DESIGNER Jack Anderson, Jana Nishi, Julia LaPine,
    Heidi Favour, Leo Raymundo
ILLUSTRATOR Julia LaPine
CLIENT Talking Rain
PRODUCT Talking Rain Sparkling Ice

The products were only sold in a warehouses, so the new packaging needed to propel them into the retail selling environment. The contemporary graphics and typography create a distinct, premium family look for the line, and the bold and playful illustration style allows for individual product differentiation.

DESIGN FIRM Hornall Anderson Design Works
ART DIRECTOR Jack Anderson
DESIGNER Jack Anderson, Julie Lock, Julie Keenan
ILLUSTRATOR John Fretz
CLIENT Starbucks Coffee Company
PRODUCT Starbuck's Mazagran four-pack/bottle

The custom-bottle shape became the icon for the Mazagran products and enhanced their shelf presence. The four-pack basket container includes a watercolor map of ancient Algeria, customized hand-lettering, and archived illustrations of the Legionnaires. The flavors are differentiated with green, gold, and black metal caps.

beverages | 135

DESIGN FIRM Sibley-Peteet
ART DIRECTOR Bryan Jesse, Rex Peteet
DESIGNER/ILLUSTRATOR Derek Welch
CLIENT Nehi/RC Cola
PRODUCT Joe Cola
TECHNIQUE Offset

This coffee beverage concept was created to take advantage of the coffee craze and the new expanded market. The design was done in Adobe Illustrator.

DESIGN FIRM Animus Comunicação
ART DIRECTOR Rique Nitzsche
DESIGNER Rique Nitzsche, Claudia Lobão
ILLUSTRATOR Roberto Renner/Lettering, João Simóes Lopes, Vinicius Cordeiro
CLIENT Fleischmann Royal
PRODUCT Royal Diet Refreshments
TECHNIQUE Offset

The Royal Company has a regular line of refreshment flavors. The new packaging is for individual servings.

DESIGN FIRM
Avalos & Bourse
ART DIRECTOR
Carlos Avalos
DESIGNER
Carlos Avalos, Allyson Brein
ILLUSTRATOR
Allyson Brein
CLIENT
SAVA Gancia
PRODUCT
Pronto Bitt
TECHNIQUE
Offset

Until a couple of years ago, the only alcoholic beverage in a can on the local market was beer. Pronto Bitt was a new concept product designed as an alcoholic ready-to-drink beverage. The consumer profile was young adults, which led to the concept for the illustration on the can.

DESIGN FIRM Avalos & Bourse
ART DIRECTOR/DESIGNER Facundo Bertranou
ILLUSTRATOR Horacio Vázquez
CLIENT Peñaflor
PRODUCT Jugos Cipolletti
TECHNIQUE Offset

Cipolletti is a premium-price regional juice, produced in an area well known for the quality of its apples. The background illustration shows a landscape of the province where these fruits are cultivated.

beverages | 137

Promotion

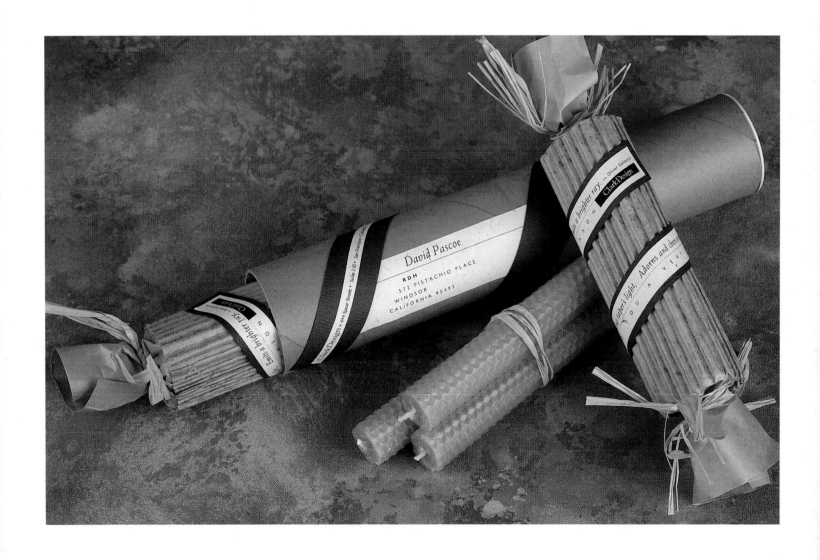

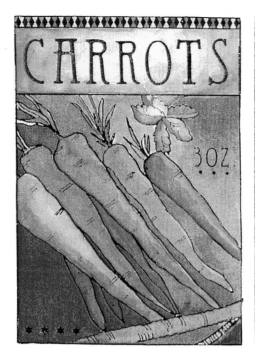

DESIGN FIRM
Red Bee Studio
DESIGNER/ILLUSTRATOR
Carole Marithe Marchese
CLIENT
Red Bee Studio
PRODUCT
Seed packets
TECHNIQUE
Gouache paintings

These seed packages were inspired by old-time fruit and vegetable labels. They were designed purely for promotional pieces and have inspired many products including T-shirts and fabric.

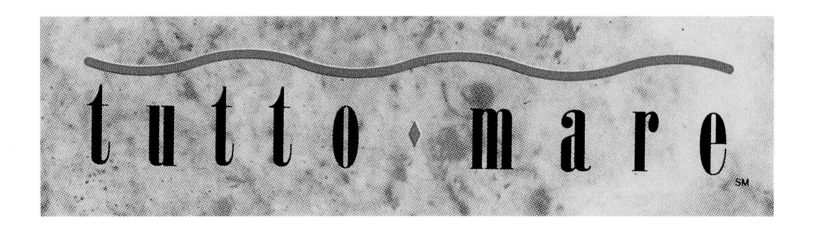

DESIGN FIRM Primo Angeli Inc.
ART DIRECTOR Primo Angeli
DESIGNER Primo Angeli, Chip Toll, Judy Radiche
CLIENT Spectrum Foods
PRODUCT Restaurant matchbox

This tiny matchbox, representing the Tutto Mare brand name of a Spectrum Foods restaurant, is the tip of the iceberg. It conveys a clear sense of the style that was carried throughout the visual system, which relied on a rich mixture of contemporary and traditional appointments.

DESIGN FIRM
Jim Lange Design
ART DIRECTOR
Genji Leclair
DESIGNER/ILLUSTRATOR
Jim Lange
CLIENT
Sears
PRODUCT
Watch
TECHNIQUE
Watch face and tin art created in Adobe Illustrator

This art was created on disk for a promotional watch, tin package, and insert card announcing the new Sears credit card design and promotional push.

DESIGN FIRM
Tieken Design and Creative Services
ART DIRECTOR
Fred E. Tieken
DESIGNER
Rik Boberg, Fred E. Tieken
CLIENT
Kiva International
PRODUCT
Promotional pizza box

Pizza slices soar above the earth, tying the graphics to the message, "Kiva customers are the best customers in the world." Sides were used as low-key mini-ads about Kiva International's service. Flying pizza slices were created in Adobe Illustrator and imported into QuarkXPress, then silk-screened on corrugated fiber pizza boxes.

promotion | 141

DESIGN FIRM Barbara Ferguson Designs
ALL DESIGN Barbara Ferguson
CLIENT Zoological Society of San Diego
PRODUCT Merchandise bags, tissue, stickers, and thank you/address card designs
TECHNIQUE Offset

The San Diego Zoo was looking for a new "fun," yet classic, look for its shopping bags. It wanted a bag that would be treasured as much as the merchandise placed in it. This unique drawing (line art created with branch twigs dipped in ink) has captured the hearts of zoo visitors.

DESIGN FIRM
Marsh, Inc.
ART DIRECTOR/DESIGNER
Greg Conyers
ILLUSTRATOR
Dan Clark
CLIENT
Kroger Stores
PRODUCT
Kroger Spice sales kit
TECHNIQUE
Offset

The objective was to design a delivery vehicle for a new spice package introduction to Kroger's sales force. The kit was very well received by the sales force and corporate offices. It was produced in Adobe Illustrator.

DESIGN FIRM Washam Design
ART DIRECTOR/DESIGNER Thurlow Washam, Linda McKenzie
CLIENT Energy Investors Funds Group
PRODUCT Collateral materials for a financial investment fund
TECHNIQUE Offset, foil stamp

The packaging needed to tie into the corporate and video look. By using video out takes and the corporate colors plus the logo, Washam Design was able to put together a package that housed all the support materials (prospectus, brochure, letter, business card, and video). The packaging was created in QuarkXPress and Adobe Illustrator.

DESIGN FIRM Curtis Design
ALL DESIGN David Curtis
CLIENT Florentine Restaurant
PRODUCT Fast-food identity system

Curtis Design created a running cartoon mascot that serves as a metaphor for fast Italian food. The energetic "Ravioli-Man" was then incorporated into a restaurant concept promoting its all-natural pasta products as a healthy and fun alternative to fried foods.

promotion | 143

DESIGN FIRM
Rickabaugh Graphics
ART DIRECTOR
Eric Rickabaugh
DESIGNER
Mark Krumel
PHOTOGRAPHER
Larry Hamill
CLIENT
Cordage Papers (paper wholesaler)
PRODUCT
Paper sample swatch books
TECHNIQUE
Offset

In an effort to tie in the natural beginnings of paper, various photos and textures were scanned. All other graphics and the package system were created in Macromedia FreeHand. Box graphics were printed on light-weight paper, laminated, and wrapped onto the cardboard shell. The swatch book covers were printed on cover stock and laminated.

DESIGN FIRM Clark Design
ART DIRECTOR Annemarie Clark
DESIGNER Craig Stout
CLIENT Clark Design
PRODUCT Holiday card/gift of beeswax candles
TECHNIQUE Xerox, hand-cut

This holiday card/gift of beeswax candles is wrapped in soft corrugated stock. A poem about the gift of light spirals around the outside of the corrugated stock. It is mailed in a tube. The design was created in QuarkXPress.

DESIGN FIRM
Clark Design
ART DIRECTOR
Annemarie Clark
DESIGNER
Thurlow Washam, Kelly Clark
ILLUSTRATOR
Kelly Clark
CLIENT
Clark Design
PRODUCT
Wine/T-shirt
TECHNIQUE
Silk screen, fiery output

These gifts were mailed to all past and present clients as a thank you for their contribution to ten successful years of business. The design was created in Adobe Illustrator; the T-shirts were silk screened, and the wine labels were hand-cut fiery output.

DESIGN FIRM Design Factory
ART DIRECTOR Amanda Brady, Stephen Kavanagh
DESIGNER Amanda Brady
CLIENT Hammet Ltd.
PRODUCT ZOUK shopping bags
TECHNIQUE Offset

These bags are printed on a paper stock with just two colors to save costs. A solid color with the logotype reversed out in white serves as a strong, yet classy, introduction to the shop, while the side panels show duotone shots of the type of wares available, and the address appears on the bottom gusset. Adobe Photoshop was used to enhance the images, Adobe Illustrator was used to create the logotype, and separations were created in QuarkXPress.

DESIGN FIRM Parham Santana Inc.
ART DIRECTOR Dean Lubensky, VH1
DESIGNER Rick Tesoro, Paula Kelly
PHOTOGRAPHER Peter Medilek
CLIENT VH1
PRODUCT Videotapes
TECHNIQUE Offset

VH1 programming was packaged as a promotion to advertisers and affiliates. Icons of the era were conceived and photographed to represent a nineties retrospective of the seventies.

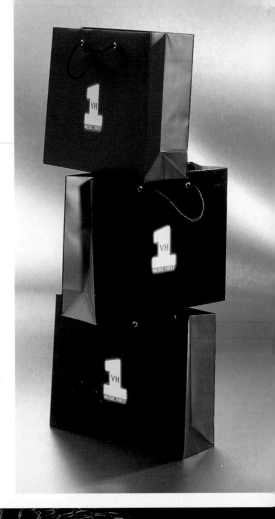

DESIGN FIRM
Parham Santana Inc.
ART DIRECTOR
Dean Lubensky, VH1
DESIGNER
Lori Reinig
CLIENT
VH1
PRODUCT
Goodie bag
TECHNIQUE
Offset

An upscale, icon-like look was conceived to house the energy of the VH1 sales presentation.

DESIGN FIRM
Mires Design
ART DIRECTOR/DESIGNER
José A. Serrano
ILLUSTRATOR
Tracy Sabin, Nancy Stahl
CLIENT
Deleo Clay Tile
PRODUCT
Clay tiles

This packaging needed to work as a shipping box as well as a point-of-purchase display. Mires Design accomplished these goals by utilizing a natural Kraft cardboard that reflected the natural quality of clay tile. It also created a die strike feature of the male figure that pops up and can function as a point-of-purchase display.

DESIGN FIRM Free-Range Chicken Ranch
ART DIRECTOR Kelli Christman
DESIGNER Toni Parmley, Kelli Christman
ILLUSTRATOR Kelli Christman
CLIENT Synopsys
PRODUCT Seminar binder
TECHNIQUE Screen

The designers selected unique material to create an innovative look and package for a two-color binder and seminar materials.

promotion | 147

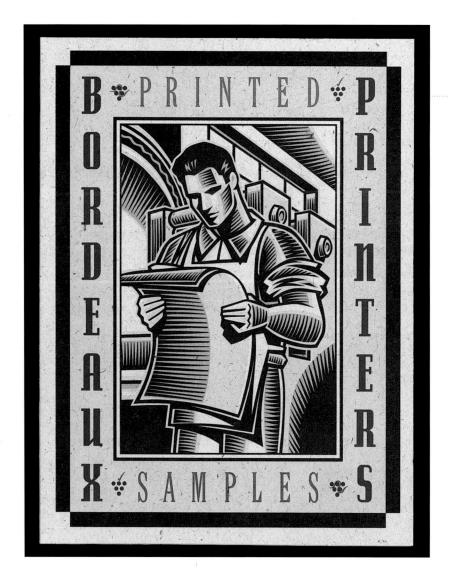

DESIGN FIRM Mires Design
CREATIVE DIRECTOR José A. Serrano
DESIGNER José A. Serrano, Miguel Perez
ILLUSTRATOR Tracy Sabin
CLIENT Bordeaux Printers
PRODUCT Bordeaux printing services

Mires Design printed labels and hang tags as part of a quality assurance program. The labels and tags were signed by sales representatives and production people to reassure clients that the printed samples had been carefully inspected.

DESIGN FIRM
Metal Studio Inc.
ART DIRECTOR
Peat Jariya
DESIGNER
Peat Jariya, Scott Head
CLIENT
Metal Studio Inc.
PRODUCT
Design images on disk and letter opener
TECHNIQUE
Offset

These labels were created in Adobe Illustrator and PageMaker. The objective was to create a system for these labels that was cost-effective, easily added to, and versatile. All are in two colors and, in some cases, interchangeable between product lines.

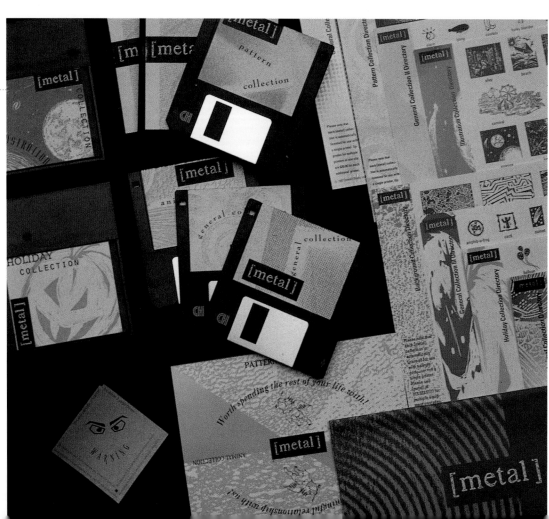

148 | Package & Label Design

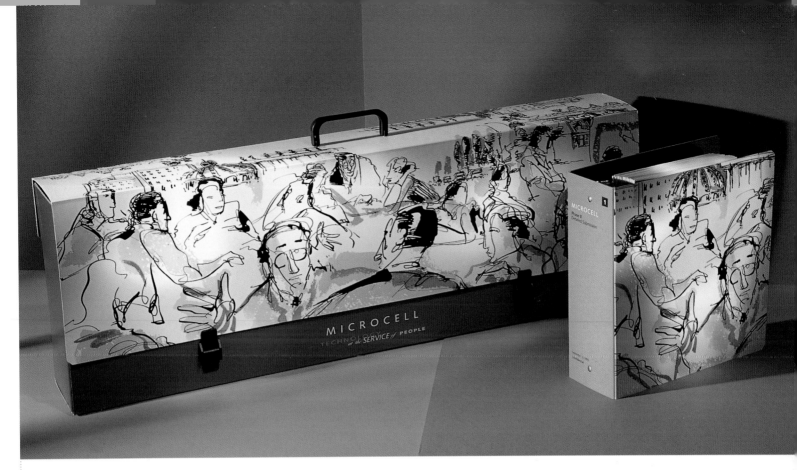

DESIGN FIRM Goodhue & Associés
ART DIRECTOR Michelle Huot, Lise Charbonneau
DESIGNER Dany Degrace, Nicole Bouchard
ILLUSTRATOR Bruce Roberts
CLIENT Microcell
PRODUCT Microcell case and binders
TECHNIQUE Screen

Enthusiasm and the innovative spirit of company personnel are conveyed through illustrations of employees' faces on binders and the presentation case.

DESIGN FIRM
Mires Design
ART DIRECTOR
José A. Serrano
DESIGNER
José A. Serrano, Deborah Hom
CLIENT
Mires Design, Inc.
PRODUCT
Self-promotion

This packaging book was created out of necessity. Mires Design received many requests for its packaging portfolio, which took a lot of time to put together. By creating this book, the firm not only saved time but also used the book as a self-promotional tool to send to clients.

promotion | 149

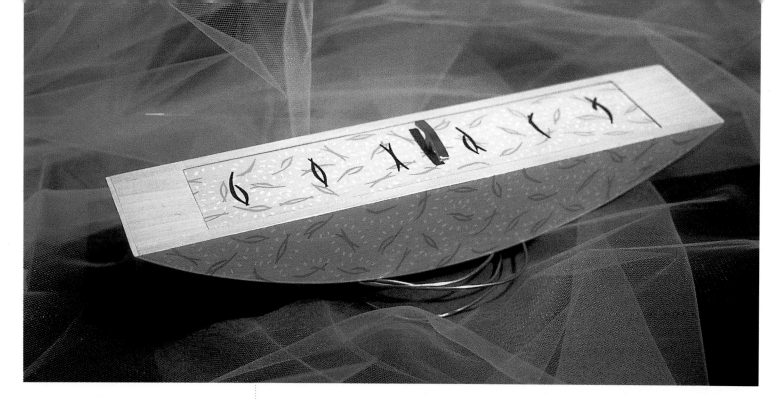

DESIGN FIRM Zubi Design
ART DIRECTOR Kristen Balouch
DESIGNER Kristen Balouch, Omid Balouch
CLIENT BoxArt Inc.
PRODUCT Christmas promotional '95
TECHNIQUE Screen

Subtly reminiscent of a monastery, this box conveys a sacred Christmas theme. Nestled inside is a pewter letter opener. "BoxArt" was embossed on the handle in the same custom-lettering that appears on the lid.

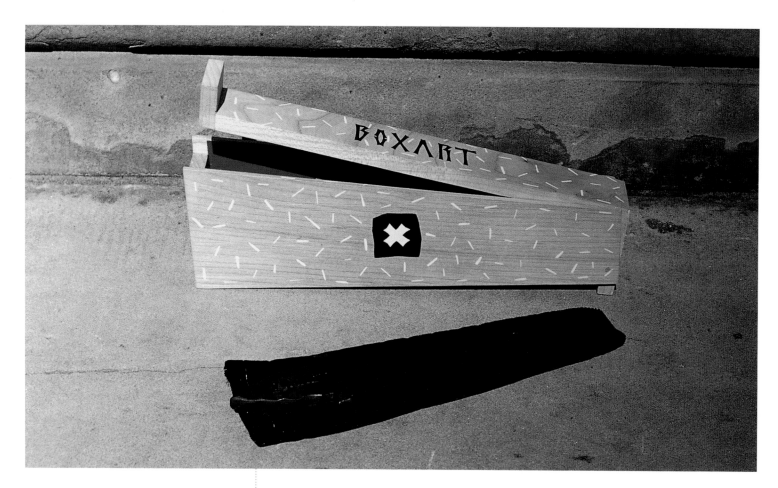

DESIGN FIRM Zubi Design
ART DIRECTOR Kristen Balouch
DESIGNER Kristen Balouch, Omid Balouch
CLIENT BoxArt Inc.
PRODUCT Christmas promotional '94
TECHNIQUE Screen

By introducing the element of instability, the rocking box challenged the conventional idea of the stable six-sided box. Packaged inside were chocolates that spelled "BoxArt" in the same custom-lettering that appears on the lid.

Package & Label Design

DESIGN FIRM Graif Design
ALL DESIGN Matt Graif
CLIENT Graif Design
PRODUCT Portfolio
TECHNIQUE Embossing, offset, color laser output

Looking for a unique way of presenting Graif Design's logos and illustrations, the designer discovered these plain looking cans and brought them to life with embossing and printing. They cost about $5 a piece to produce plus time to assemble.

DESIGN FIRM Graif Design
ALL DESIGN Matt Graif
CLIENT Graif Design
PRODUCT Beer
TECHNIQUE Embossing on silver metallic paper

The designer makes his own beer, so he created this package to showcase his package design ability. The entire piece was created in Adobe Illustrator. Labels are hand die-cut and applied. They cost $8 each to produce plus time to assemble.

DESIGN FIRM
R. Scholbrock Works
DESIGNER/ILLUSTRATOR
Roger Scholbrock
CLIENT
R. Scholbrock Works
PRODUCT
Brown Ale #2 (home brew)
TECHNIQUE
Color ink jet printer

This product is part of a series of home brewed beer that R. Scholbrock Works gives clients in appreciation for their business. The label was produced in Macromedia FreeHand.

DESIGN FIRM
Westpoint Stevens, Inc. design group
ART DIRECTOR/DESIGNER
Gail Rigelhaupt
CLIENT
Westpoint Stevens, Inc.
PRODUCT
Wine
TECHNIQUE
Offset

This specially packaged red wine was given as a promotional and "image" gift to the press, sales force, and selected retailers to connect the bedding with a "vintage" product. The label was developed to showcase the company's new "vintage" bedding collection. It named the blend "Rosetta" after the pattern from which it was designed.

152 | Package & Label Design

DESIGN FIRM Love Packaging Group
ALL DESIGN Tracy Holdeman
CLIENT Love Box Company
PRODUCT Trade show carry-all box
TECHNIQUE Flexography on corrugated board

Because Love Box Co. is a corrugated box manufacturer, it designed the box to be printed flexographically with three colors on corrugated board. The design was created by hand except the type, which was set in Macromedia FreeHand and output on a 600 dpi laser printer. The colors were chosen so that they would work when overprinted and therefore involve no trapping.

DESIGN FIRM Sayles Graphic Design
ALL DESIGN John Sayles
CLIENT James River Paper Corporation
PRODUCT Curtis Brightwater
TECHNIQUE Offset, screen

Curtis Brightwater papers were introduced with a splash. John Sayles developed a unique presentation inspired by the product's name—bottles of "brightwater" accompany swatch books in a beverage carrier.

promotion | 153

DESIGN FIRM Sayles Graphic Design
ALL DESIGN John Sayles
CLIENT National Travelers Life
PRODUCT President's Club cruise clock
TECHNIQUE Screen

NTL commissioned Sayles to create a keepsake to commemorate the company's annual cruise for top producers. The small quantity (only 75 were produced) allowed Sayles to develop a spectacular hand-assembled clock made from a variety of materials. Individual pieces of the clock are mounted on a base of dark blue Avonite. Layers of Plexiglas—translucent orange and opaque white—form a horizon behind the aluminum ship, while copper wire creates sun rays around the clock's face. The clock is presented in a box adorned with nautical graphics in copper and blue.

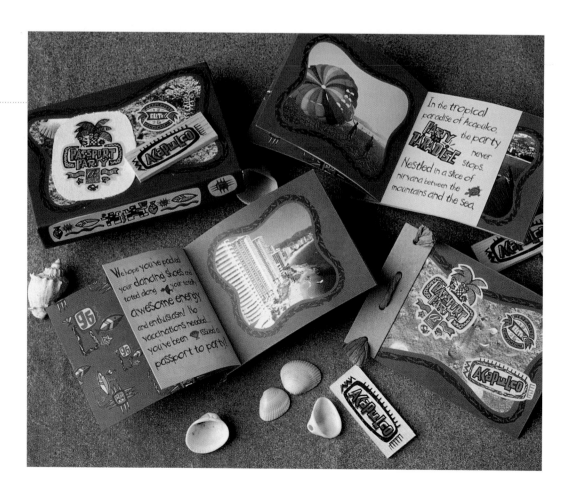

DESIGN FIRM
Sayles Graphic Design
ART DIRECTOR/DESIGNER
John Sayles
ILLUSTRATOR
John Sayles, Jennifer Elliott
CLIENT
Bally Total Fitness
PRODUCT
Passport to Party promotion
TECHNIQUE
Offset

The fitness company recognizes top performers with a trip outlined in a multipage brochure designed by Sayles Graphic Design. The piece arrives in a graphic, hinge-lid box adorned with screen-printed wood and canvas. The brochure is bound with Kraft raffia. The inside pages are different sizes and colors, printed with visuals that complement the box.

154 | Package & Label Design

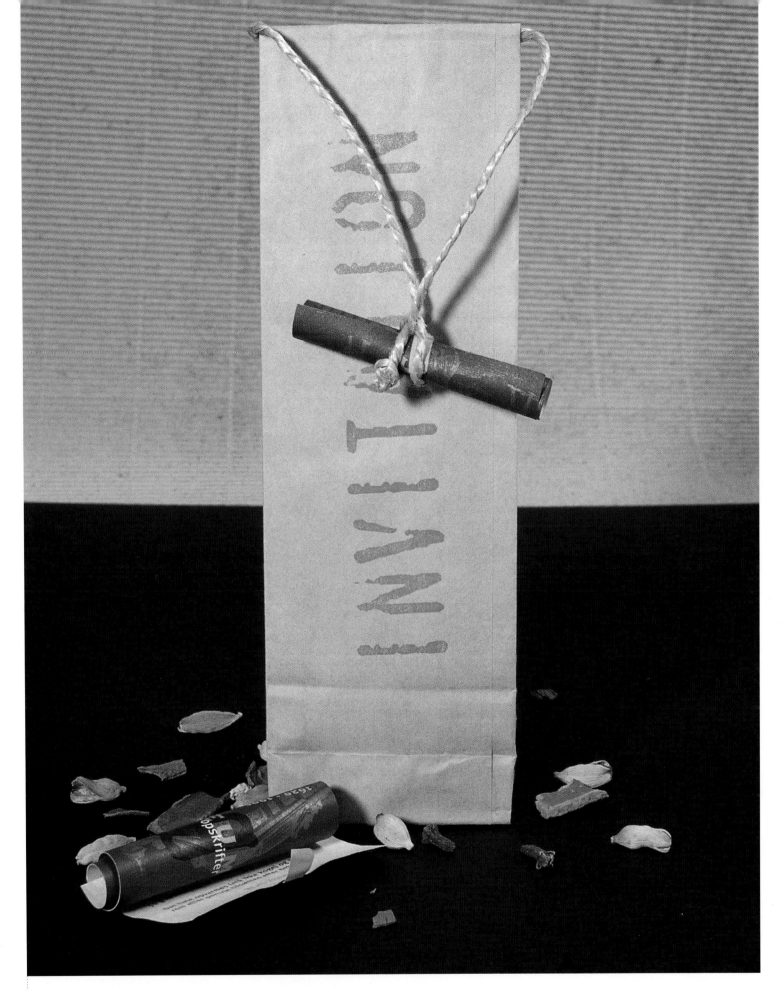

DESIGN FIRM Leo Pharmaceutical, in-house
ART DIRECTOR/DESIGNER Vibeke Nødskov
CLIENT Leo Pharmaceutical
PRODUCT Glogg Mix (herbs and spices)
TECHNIQUE Bag–silk screen, handle–offset

The brown paper bag works as an invitation to an exhibition. The guest is encouraged to bring along the bag and have it filled at the exhibition with Glogg Mix, a hot spicy wine. A recipe is folded and rolled as a handle with a little orange sticker and a piece of string so that the bag can be carried around and taken home for preparation.

DESIGN FIRM
Design Ahead
DESIGNER
Ralf Stumpf
CLIENT
Design Ahead
PRODUCT
Self-promotion

This box, filled with the Design Ahead Gearwheel logo made of marzipan, was sent to special clients as a Christmas gift.

DESIGN FIRM
K-Products
ALL DESIGN
John Vander Stelt
CLIENT
Self-promotion
PRODUCT
Outbreak

This promotion introduced K-Products' new Outbreak Jacket. It was mailed to 378 buyers, and fifty-two placed orders for the new jackets. Details such as the pine scent pouch, colorful fall leaf, and brown tissue paper make this a memorable promotion.

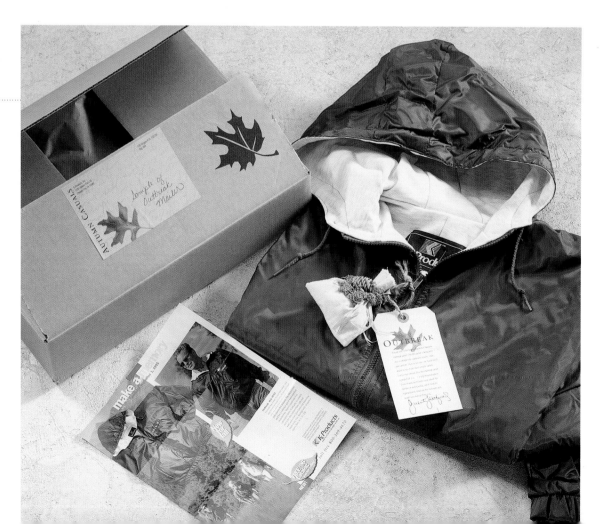

156 | Package & Label Design

DESIGN FIRM Greteman Group
ART DIRECTOR/DESIGNER James Strange
ILLUSTRATOR Bill Gardner
CLIENT Menefee & Partners
PRODUCT Packaging

This Pebble Beach promotion uses real wooden boxes with burnt covers. A regional illustration of the California wine country and color captures the proper look. It was created in Macromedia FreeHand.

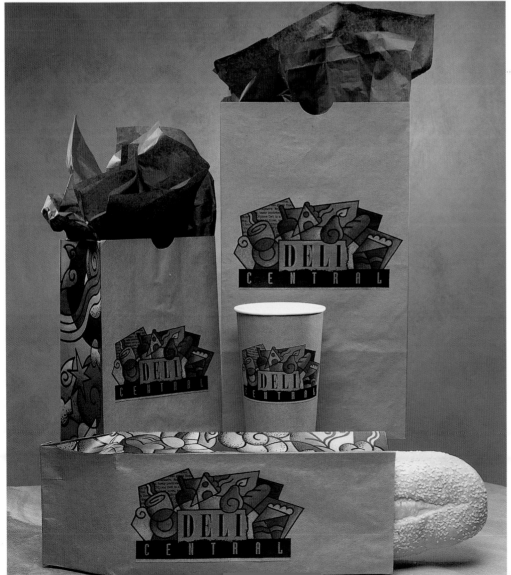

DESIGN FIRM Greteman Group
ART DIRECTOR Sonia Greteman
DESIGNER Sonia Greteman, James Strange
ILLUSTRATOR Sonia Greteman
CLIENT Dillons Food Stores
PRODUCT Grocery bags

These Deli Central sacks and cups were designed to create an upscale European feel in a section of a traditional grocery store. The logo and art have a cubist flavor, using icons of the foods offered at the deli. It was created in Macromedia FreeHand.

DESIGN FIRM Animus Comunicação
ART DIRECTOR Rique Nitzsche
DESIGNER Felicio Torres
CLIENT Animus Comunicação
PRODUCT Christmas gift
TECHNIQUE Inkjet

This Christmas promotional gift was designed on the computer and printed on an inkjet color printer. The labels were hand-located by the Animus Comunicação staff. All of the labels were personalized with the name of the person who received the gift.

158 | Package & Label Design

Index

Able Design   50, 51
54 West 21st Street, Suite 705
New York, NY 10010

Acme Creative Group   23, 24
1515 Broadway, 39th Floor
New York, NY 10036

Adam Cohen Studio   10
252 W. 17th Street
New York, NY 10014

After Hours Creative   33
1201 East Jefferson B100
Phoenix, AZ 85034

Angelo Sganzerla   56, 80, 81, 82
Via Crema 27
Milano 20135
Italy

Animus Comunicação   40, 96, 97, 136, 158
Ladeira do Ascurra, 115-A
Rio de Janeiro, RJ 22241-320
Brazil

Avalos & Bourse   137
La Pampa 1351
Buenos Aires 1428
Brazil

Barbara Ferguson Designs   112, 142
10211 Swanton Drive
Santee, CA 92071

Bartels & Company, Inc.   91
3284 Ivanhoe Avenue
St. Louis, MO 63139

Boulder Design & Illustration   68
283 South Union, Unit #4
Burlington, VT 05401

Bunny Levy & Associates   131
2A Ruthven Road
Kingston 10
Jamaica WI

Carmichael Lynch   26, 80
800 Hennepin
Minneapolis, MN 55403

Carter Wong and Partners Ltd.   107
29 Brook Mews North
London WZ 3BW
England

Chiron Diagnostics   17
63 North Street
Medfield, MA 02052

Clark Design   18, 19, 120, 144, 145
444 Spear Street, #210
San Francisco, CA 94105

CMA   22, 77, 78, 100, 121, 122
1207 Dunlavy
Houston, TX 77019

The Coleman Group   75, 116
305 East 46th Street
New York, NY 10017

CommuniQué Marketing   16, 32
1520 West Main Street, Suite 106
Richmond, VA 23220

Curtis Design   51, 73, 143
3328 Steiner Street
San Francisco, CA 94123

Damion Hickman Design   69
1801 Dove, #104
Newport Beach, CA 92660

Daniel K. Brown   114
c/o Wickham
4636 Pine Harrier Drive
Sarasota, FL 34231

Del Rio Diseño   73, 94, 113
Nueva Los Leones, #0233 D/F
Santiago
Chile

Desgrippes Gobé & Associates   62, 63
411 Lafayette Street
New York, NY 10003

Design Ahead   156
Kirchfeldstr. 16
45219 Essen-Kettwig
Germany

The Design Company   60, 89, 132
One Baltimore Place, Suite 170
Atlanta, GA 30308

Design Communications   13
1333 North Kingsbury Street
Chicago, IL 60622

Design Factory   145
3+4 Merrion Place
Co. Dublin
Ireland

Design Resources   118
1955 Cliff Valley Way, #230
Atlanta, GA 30329

Design Source, Inc.   118
R & D Building, Room 204
7615 Guijo Street
San Antonio Village, Makati City
Philippines

DesignCentre of Cincinnati   57, 126
225 East Sixth Street
Cincinnati, OH 45202

DIL Consultants in Design   120
R. Oscar Freire, 379–conj. 162
São Paulo, SP–Brazil 01426-001

FLB Design Limited   87, 88, 128, 129
De La Bere House, Bayshill Road
Cheltenham GL50 3AW
England

fotodesign + conception   110
Pfarrius strasse 12
D-50935 Koeln
Germany

FRCH Design Worldwide   46, 108
444 North Front Street, #211
Columbus, OH 43215

Free-Range Chicken Ranch   147
330A East Campbell
Campbell, CA 95008

Goodhue & Associés   42, 43, 98, 99, 149
465 McGill Street, 8th Floor
Montréal, Québec H2Y 1H1 9
Canada

Graef & Ziller Design   37
330 Fell Street
San Francisco, CA 94102

Graif Design   151
985 South Firefly Road
Nixa, MO 65714

Greteman Group   33, 61, 92, 134, 157
142 North Mosley
Wichita, KS 67202

Haley Johnson Design Company   47, 70
3107 East 42nd Street
Minneapolis, MN 55406

Hans Flink Design   52, 117
224 E. 50th Street
New York, NY 10022

Heather Sumners   123
William Carey College
1836 Beach Boulevard, Box 59
Gulfport, MS 39507

The Hive Design Studio   79
126 Plum Street
Santa Cruz, CA 95062

Hornall Anderson Design Works   30, 38, 92, 93, 135
1008 Western Avenue, Suite 600
Seattle, WA 98104

jCHO   123
815 Buena Vista
Pascagoula, MS 39568

Jeff Labbé Design Company/ dGWB Advertising   49
1764 Mason Street
San Francisco, CA 94133

Jim Lange Design   141
203 North Wabash Avenue, #1312
Chicago, IL 60601

K-Products   156
1520 Albany Place Southeast
Orange City, IA 51041

Karacters Design Group   129, 130
1700-777 Hornby Street
Vancouver, BC V6Z 2T3
Canada

Lambert Design   74, 95
7007 Twin Hills Avenue, Suite 213
Dallas, TX 75231

Index

Leo Pharmaceutical 155
Industriparken 55
2750 Ballerup
Denmark

Lewis Moberly 37, 50, 90, 132
33 Gresse Street
London WIP 2LP
England

Louisa Sugar Design 52
1650 Jackson Street, Suite 307
San Francisco, CA 94109

Louise Fili Ltd. 20, 21, 54, 76, 77, 120
71 Fifth Avenue
New York, NY 10003

Love Packaging Group 34, 36, 153
410 East 37th Street North
Plant 2, Graphics Department
Wichita, KS 67219

Maddock Douglas, Inc. 83
257 North West Avenue, Suite 201
Elmhurst, IL 60126

Malik Design 60, 61
88 Merritt Avenue
Sayreville, NJ 08879

Mars Advertising 11
24209 North Western
Southfield, MI 48075

Marsh, Inc. 15, 114, 142
34 West 6th Street, #1100
Cincinnati, OH 45202

Martin Ross Design 41
1125 Xerxes Avenue South
Minneapolis, MN 55405

Metal Studio Inc. 148
1210 West Clay, Suite 13
Houston, TX 77019

Mike Salisbury Communications, Inc. 101, 126
2200 Amapola Court, Suite 202
Torrance, CA 90501

Minkus & Associates 69
100 Chetwynd Drive, Suite 200
Rosemont, PA 19010

Mires Design 28, 29, 30, 31, 58, 127, 147, 148, 149
2345 Kettner Boulevard
San Diego, CA 92101

Morla Design 53
463 Bryant Street
San Francisco, CA 94107

One Design 17
P.O. Box 419
13 Fox Lane
Quechee, VT 05059

Parham Santana Inc. 25, 146, 147
7 West 18th Street
New York, NY 10011

Paul Ng Design & Productions 75
80 Acadia Avenue, Suite 302
Markham, ON L3R9V1
Canada

Pedersen Gesk 27, 28, 84, 85
105 5th Avenue South, Suite 513
Minneapolis, MN 55401

Pentagram Design Inc. 18, 53, 115
204 Fifth Avenue
New York, NY 10010

Pinkhaus 100
2424 South Dixie Highway
Miami, FL 33133

Planet Design Company 39, 65
605 Williamson Street
Madison, WI 53703

Primo Angeli Inc. 70, 104, 105, 106, 140
590 Folsom Street
San Francisco, CA 94105

R. Scholbrock Works 152
115 North Main Street
Potosi, WI 53820

Red Bee Studio 140
77 Lyons Plain Road
Weston, CT 06883

revoLUZion 59, 87
Uhlandstrasse 4
D-78579 Neuhausen ob Eck
Germany

Richards Group Inc. 91
375 Thomas More Parkway
Crestview Hills, KY 41017

Rickabaugh Graphics 22, 54, 144
384 West Johnstown Road
Gahanna, OH 43230

Robert Bailey Incorporated 125
0121 Southwest Bancroft Street
Portland, OR 97201

Robilant & Associati 23, 55, 124
Via Vigevano 41
20144 Milano
Italy

Rocha & Yamasaki 35, 59, 86
Rua Pe. João Manuel
1078, Casa 6
São Paulo, SP 01411-000
Brazil

Saatchi & Saatchi Denmark 133
Overgaden Oven Vandet 54 A 2
DK-1415 Copenhagen
Denmark

Sayles Graphic Design 35, 90, 133, 153, 154
308 Eighth Street
Des Moines, Iowa 50309

Shimokochi/Reeves 48, 72
4465 Wilshire Boulevard
Los Angeles, CA 90010

Sibley-Peteet Design 38, 39, 63, 64, 94, 136
3232 Mckinney, Suite 1200
Dallas, TX 75204

Smart Design Inc. 16
137 Varick Street, 8th Floor
New York, NY 10013

Sommese Design 110
481 Glenn Road
State College, PA 16803

Steep 124
Box 89
South Glastonbury, CT 06073

Steve Trapero Design 31
3309-G Hampton Point Drive
Silver Spring, MD 20904

Stoltze Design 24
49 Melcher Street, 4th Floor
Boston, MA 02210

Studio Associates 116
5850 Chase Road
Dearborn, MI 48126

Susan Healy Design 85
3018 Sunnynook Drive
Los Angeles, CA 90039

Tangram Strategic Design 36
Via Negroni 2
Novara 28100
Italy

Teikna 55
366 Adelaide Street E #541
Toronto, ON M5A 3X9
Canada

Tharp Did It 109
50 University Avenue, Suite 21
Los Gatos, CA 95030

Tieken Design and Creative Services 107, 109, 141
2800 North Central Avenue, Suite 150
Phoenix, AZ 85004

Vanessa Eckstein 57, 58
430 South Cloverdale Avenue, #7
Los Angeles, CA 90036

Washam Design 119, 143
444 Spear Street, #210
San Francisco, CA 94105

Watts Graphic Design 12, 40, 71, 111
79-81 Palmerston Crescent
South Melbourne, VIC 3205
Australia

Westpoint Stevens Inc. 128, 152
1185 Avenue of the Americas
New York, NY 10036

Wood & Wood 13, 14
101 North Scenic Hills Circle
North Salt Lake, UT 84054

Zubi Design 150
57 Norman Avenue, #412
Brooklyn, NY 11222